almost human

also by Lee Gutkind

IN FACT: THE BEST OF CREATIVE NONFICTION

FOREVER FAT: ESSAYS BY THE GODFATHER

THE VETERINARIAN'S TOUCH

*STUCK IN TIME: THE TRAGEDY OF
    CHILDHOOD MENTAL ILLNESS*

*CREATIVE NONFICTION:
    HOW TO LIVE AND WRITE IT*

ONE CHILDREN'S PLACE

*MANY SLEEPLESS NIGHTS: THE WORLD OF
    ORGAN TRANSPLANTATION*

THE PEOPLE OF PENN'S WOODS WEST

*THE BEST SEAT IN BASEBALL,
    BUT YOU HAVE TO STAND!*

GOD'S HELICOPTER (A NOVEL)

BIKE FEVER

almost human

MAKING ROBOTS THINK

Lee Gutkind

W. W. NORTON & COMPANY

NEW YORK LONDON

For information about permission to reproduce
selections from this book, write to Permissions,
W. W. Norton & Company, Inc.
500 Fifth Avenue, New York, NY 10110

Manufacturing by Quebecor Fairfield
Book design by Barbara M. Bachman
Production manager: Julia Druskin

LIBRARY OF CONGRESS CATALOGING-IN-PUBLICATION DATA

Gutkind, Lee.
 Almost human : making robots think / Lee Gutkind.
    p. cm.
  Includes biographical references.
  ISBN-13: 978-0-393-05867-3 (hardcover)
  ISBN-10: 0-393-05867-0 (hardcover)
 1. Autonomous robots. 2. Artificial intelligence. I. Title.
  TJ211.495.G88 2007
  629. 8'9263—dc22

                                        2006101046

W. W. Norton & Company, Inc.
500 Fifth Avenue, New York, N.Y. 10110
www.wwnorton.com

W. W. Norton & Company Ltd.
Castle House, 75/76 Wells Street, London W1T 3QT

1 2 3 4 5 6 7 8 9 0

This book is dedicated to my son,
    Sam Gutkind,
whose love and loyalty, faith and presence,
inspire and enrich me.

contents

Introduction: The Rookie Revolution   *xi*

PART ONE: the atacama

    1. Wild Ride to Base Camp   *3*

    2. Big Red   *7*

    3. Groundhog   *24*

    4. Bummed and Elated   *35*

    5. A Lack of Vision   *48*

PART TWO: autonomy

    6. RoboCup   *61*

    7. The Color of Thinking   *78*

    8. Asimo and Friends   *95*

    9. Frustration   *100*

    10. The Challenge   *120*

    11. Fresh Blood   *131*

PART THREE: the ops

    12. The Grasshopper and the Ant   *145*

    13. Fallback Positions   *162*

14.   The Desert Makes Us Wacky   *176*

15.   Peeing on a Rock   *193*

16.   Downtime   *206*

17.   Two Versions of Reality   *219*

PART FOUR: making history

18.   Nathalie   *227*

19.   Pirate's Cove   *237*

20.   Hardware vs. Software   *255*

21.   In the Field   *265*

22.   The Barest Beginning   *270*

Acknowledgments   *281*

Notes   *283*

The Rookie Revolution

OVER THE PAST HALF-DOZEN YEARS, OFF AND ON, I have been a fly on the wall at the Carnegie Mellon Robotics Institute in Pittsburgh, witness to a vital new movement coming to life.

What has happened at Carnegie Mellon's Robotics Institute during that period changed the shape and scope of the way in which robot technology has evolved. The roboticists at Carnegie Mellon have not only created cutting edge ideas and applications for robots, but they have also embraced and brought together the opposing factions in the robotics world—the engineers who build the bodies of the robots and the software specialists, the code writers, who help make the robots think. "Code" means directions written in programming language that intricately translate a robot's every conceivable action.

We can credit this accommodation or truce between the hardware/software factions, and the great advances in technology it fostered, partially to Institute leadership. It is a loosely structured operation, with faculty and staff accorded great flexibility and independence—an atmosphere that fosters creative freedom and encourages synergistic collaboration within the institution and well beyond it.

But what makes the Robotics Institute unique and, I believe, the reason why the robotics movement is at the tipping point now, is that the proletariat—the students—share equally in the process. Students are a strong and vital force behind the creation of the robots produced at Carnegie Mellon. The amazing robots you will meet in this book are primarily the products of the brain and brawn of men and a few women under thirty years of age.

While the technological concepts necessary to make robots think and act are highly sophisticated, a very traditional apprenticeship system is in place in which students or former students learn through on-the-job, hands-on experiences. Young people—they are jokingly referred to as "rookies" or "fresh blood"—don't just do the dirty work, they experience all aspects of the process. There are, as you will see, astounding benefits, along with unfortunate drawbacks, to launching and fortifying a revolution with rookies.

But the robotics world would not be at the tipping point, and we would not be contemplating and initiating interplanetary travel with robots, or robots that compete with humans in soccer, or robots that do science on Mars, or robots that treat patients in hospitals, or robots that engage in realistic conversation with anyone who chooses to speak to them, without the vitality and incredible creativity of the new generation. In the 1990s, computers and the evolving technology— from cell phones to video games—changed the world because of the passion and persistence of young people. Robotics energized by rookies will transform technology again in the twenty-first century.

At Carnegie Mellon graduate students run the gauntlet of evaluation by simply being accepted into this very competitive program, and then allowing themselves to be tossed into the fire and brimstone of the robot tidal wave that obsesses most everyone on the premises. Once you are part of this vortex, the need to succeed, the drive to reach a technological milestone, becomes relentless. The frenetic pace of graduate life is numbing but addictive. Recently, Matt Mason, director of the Robotics Institute, observed of his grad students: "They are smart, but naïve; you have to be careful not to tell them that the things we are expecting them to do are impossible."

As you will see, Mason wasn't joking. While robots will not soon assume a significant role in society, the advances achieved over the past decade have gone far beyond the expectations of all but the wildest dreamers. Students and freshly minted graduates are the main engines of momentum. *Almost Human: Making Robots Think* demonstrates how and why this is happening. The future of robotic technology and the power and influence it will yield, is overall, I am pleased to report, in good hands.

PART ONE
the atacama

CHAPTER ONE

Wild Ride to Base Camp

Francisco Calderón, the Chilean student and translater whom everyone calls "Finch," is waiting for us at the entrance to the tiny airport in Iquique. Of my traveling companions, Alan Waggoner and Paul Tompkins are veterans of previous Atacama expeditions, while Dom Jonak and I are viewing this eerie landscape for the first time. Jonak and Tompkins are from the Robotics Institute at Carnegie Mellon University in Pittsburgh while Waggoner directs the Molecular Biosensor and Imaging Center (MBIC) at Carnegie Mellon.

The Atacama Desert stretches from the Peruvian border south in a narrow band 600 miles into northern Chile. With its dazzling white salt flats and vast expanses of rusty-red emptiness, the Atacama is the driest place on Earth, a place climatologists call absolute desert. Death Valley in California and the Gobi Desert in Mongolia get anywhere from three to six times more moisture annually than the Atacama. Obviously, there are very few living organisms in this desert, which makes the Atacama an ideal analog to Mars. Seeking life on Mars is an ongoing obsession of many of the scientists and software engineers at Carnegie Mellon and the National Aeronautics and Space Administration (NASA), a frequent partner.

We pile our bags into the bed of the double cab Toyota HiLux pickup, a sturdier version of the Toyota Tacoma we use in the United States. The HiLux, an Action Utility Vehicle (AUV) according to Toyota, is aptly named because of its capacity to endure constant, violent, and aggressive abuse. They are perfect for the terrain and also for the crazed, explosive spirit with which the roboticists drive when unleashed in the wilds with a vehicle built for battering. The programmers' frustration of sitting behind a computer and writing and struggling endlessly with code is released in this desert—with passion. Soon, we are rocketing out onto this sun-scorched plain.

The road from the airport into the desert is smooth, and the terrain is flat and red, with white misshapen clumps of salt dotting the landscape to the west where we are headed. They look like gigantic white fat globules, glittering, almost oozing, in the sun. We can see the distant bejeweled reflection of the ocean off to one side, a rather disconcerting sight, considering the barrenness of the landscape surrounding us. Although it does not form rain, the moisture contained in the thick dense fog from the ocean seeping above the mountains allows some organisms to survive. We will be seeking those organisms at Salar Grande, the location of our base camp, less than an hour away. Iquique—the name comes from the Aymara (a native Andean ethnic group) word that translates to "laziness"—is a tourist stop popular for its surf and beaches and its architecture. There's also a bustling commercial port area to service the copper and salt unearthed from mines in the desert.

The road we're on ends after a high-speed half-hour of full-throttle straightaway driving. Finch—so called because he resembles the character of the same name in the movie *American Pie*—skids to a stop in front of the security gate of the salt mine, the Compania Minera Punta de Lobos, the biggest open cast mine of common salt in the world. Swarthy, curly haired, and slender, Finch jumps out of the truck, places a flashing yellow light on top of the cab, and releases a high-flying antenna, which has been secured against the cab, with a red flag at its tip. This is to make us as visible as possible to avoid dangerous collisions with the gigantic tractor-trailer salt trucks, which commute from the mine to the port where the salt ships are loaded for transport, to spots around the world for refining.

At this point, leaving the mine and the ocean behind us, the smooth road peters out and in its place is a rutted washboard rock-and-dirt right-of-way that has not been maintained for many years, if ever. There's a single deep set of tire tracks seared into the landscape and extending into a red, dust-swirling infinity. "Hang on," Finch says in his heavily accented English, turning to look at us, and smiling. "To minimize brain damage, I have to drive like a bat out of hell. We will soar over the bumps."

This then is our brief introduction to the most harrowing and frenzied aspect of desert life during field operations, called OPS: Driving with a roboticist. The HiLux has a roll bar, I am happy to note.

A few weeks later, one of these trucks, rocketing down this road at breakneck speed while racing with another in the pitch dark dead of night, would flip over, sumersaulting its occupants, breaking windows and collapsing metal. Luckily, the young roboticists in the cab will survive without too many scratches.

Today, ten minutes of wild, bumping, swerving, back-breaking, jaw-crunching, roller-coastering later, with our backs aching and our knees bruised, we are all about to puke our guts out, when, at an invisible marker, Finch suddenly takes a wild turn right, skids on the salt-laden sand like a downhill skier, and streaks up a steep hillside in an explosion of red dust. The base camp is on top of the hill.

Zoë is on the periphery as we pull in. It is wide and low to the ground, sitting on thick mountain bike tires. A flat row of shiny solar panels energizes a row of batteries below the panels. Three high-resolution digital cameras gawk into the bright sun atop a scrawny cranelike neck. It reminds me of an ice cream cart at a carnival—or a self-propelled flatbed railroad car. Zoë's watching as we stumble out of the truck. Or at least Zoë *seems* to be watching. We don't know at the time that Zoë can't see. Or, to put it more precisely, Zoë can see, technically; but what Zoë sees isn't exactly what is there—a fact that makes the prospects for a successful OPS very tenuous.

CHAPTER TWO

Big Red

Zoë, THE WORD FOR "LIFE" IN GREEK, IS PART OF A three-year, $4 million project called "Life in the Atacama" or LITA for short, funded by NASA's ASTEP (Astrobiology, Science, Technology Exploration of Planets) program. Astrobiology is the study of the origins, evolution, and future of life in the universe. ASTEP was established "to enable future space missions to determine whether life exists or has existed outside Earth," according to NASA. This will be done not through efforts in outer space, however, but "through a detailed, collaborative analysis of Earth's extreme environments, so we can better prepare to understand analogous systems elsewhere." This in a nutshell is the primary objective of LITA.

While sending men to Mars or other planets may someday be possible, a prevailing point of view at NASA and many other research organizations is that robots may lead the way. The question is: Considering the hardware and software available today, what can a robot accomplish? Can a robot navigate the rough Martian terrain and go long distances *on its own* without getting bogged down in sand or disabled by rock and ravine? Humans won't be around for emergency rescue operations.

More important, will a robot be capable of conducting scientific experiments about the geology and biology of the planet, thereby justifying the significant investment necessary to build and transport a robot so many millions of miles from Earth? Zoë is not intended to be *the* robot that will end up on Mars, by the way; it is, rather, a "terrestrial test bed"—this is NASA terminology—a way to telescope into the future to see if autonomy of locomotion and scientific performance is possible.

If Zoë is successful traversing the Atacama in an autonomous mode (without human interference) and if it can perform scientific experiments (autonomous science) the value and potential of the robot scientist will be established. Then the Robotics Institute at Carnegie Mellon will be at the leading edge of the intersection of science and technology. For Carnegie Mellon specifically, and the future prospects of robots as exploratory instruments, a great deal was at stake with LITA.

The upcoming OPS are significant. The OPS will test Zoë in an on-the-job "Mars-relevant" work environment, meaning that its mission in the Atacama will simulate a rover similar to Zoë landing on Mars. Over the next month, Zoë will be expected to function in the field while being directed by scientists and roboticists thousands of miles away, a process similar to how a real Mars mission might proceed. This is the first of two opportunities in 2004 for Zoë to demonstrate what it can do. The goal this year is to do enough testing so that the hardware and software elements can be analyzed, modified and integrated, and made ready for a flawless and successful final OPS a year away. While the 2005 OPS will be documented in the scientists and roboticists' journal publication papers and in reports to NASA, the 2004 OPS is in many ways more important. It will determine the extent of the potential of the technology that has been developed for Zoë.

No other organization or institution is more perfectly suited for this conceptual test bed role. Carnegie Mellon's Robotics Institute has created many large "field" (automated work machines for land, sea, air, and space) robots over the previous three decades. In 1983, the Remote Reconnaissance Vehicle (RRV), a clunky six-wheeled creature, had entered the contaminated reactors at Three-Mile Island power plant in Pennsylvania, which had suffered a partial meltdown in 1979, to inspect the sealed-off radiation danger zone and clean up the flooded areas. A sister robot, the CoreSampler, subsequently drilled core samples from the walls to determine the depth and severity of radioactive material soaked into the concrete.

Carnegie Mellon's William "Red" Whittaker, the "father of field robotics," was, along with Raj Reddy, the visionary at the Robotics Institute back then. Buoyed by the success of the RRV and CoreSampler, Whittaker helped create other Carnegie Mellon field robots, including Terregator, the world's first outdoor navigation robot, and Rex, the world's first autonomous digging machine. Another visionary, a grad student named Chuck Thorpe, created NavLab, a van with lasers and computers, which, after ten years in development, completed a Washington, D.C., to Los Angeles "No Hands Across America" (researchers handled the throttle and brake) highway journey in 1995. At the same time NavLab was started, Whittaker developed the Remote Work Vehicle, popularly known as Workhorse, was developed as an overall clean-up robot—a commercial product—to wash contaminated surfaces, remove sediments, demolish radiated structures, and transport materials.

The Robotics Institute was established in 1979 as part of the University's Computer Science Department, which was founded by Nobel Prize winner Herbert A. Simon, the father of artificial intelligence, and Allen Newell, a pioneer in the use of computer simulation for understanding and modeling the human mind. Raj Reddy, who was later to become dean of the School of Computer Science, founded the Robotics Institute by persuading the Westinghouse Electric Corporation to part with a $5 million grant for students to work at the Institute on Westinghouse projects.

Today nearly all of the Institute—which includes 75 full-time and adjunct faculty members, 15 postdoctoral fellows, nearly 150 PhD and master's degree students, and as many as 300 visitors, researchers, and staff members—works on government-funded projects, primarily from the DoD (Department of Defense), NASA, and NSF (the National Science Foundation). Westinghouse has faded from Pittsburgh's economic landscape, as have most of the international corporations once headquartered there. The major universities—Carnegie Mellon and the University of Pittsburgh—are today the chief engines of growth in the metropolitan area.

Reddy's timing was impeccable, especially with a true believer like Whittaker in the fold. Whittaker was intense and intimidating. In meetings he would snap ballpoint pens in his hands if discussions were not going his way. Whittaker watchers—and there have always been thousands in Pittsburgh, fascinated by his charismatic energy—remember the young Whittaker as a wild-eyed giant, six feet four inches tall, nearly 250 pounds, with long red hair, and Levi's so torn and mangled you could, a female colleague observed, "see almost everything." He was a mountain climber, a competitive boxer, a long-distance runner, a notable presence on campus, regularly returning from a long run in the middle of winter, half-naked, "with icicles dripping from his beard," according to a close friend. Whittaker, who claims to have wrestled a gorilla in high school on a dare, once told a reporter: "There was a time in my life when I would just as soon break your face as look at you." He was then and is now oblivious to personal details. Hotels ship home luggage because he neglects to return to his room. He will shuttle to his office from the airport, forgetting his car in the parking lot. He instinctively drains his brain of extraneous details, while absorbing and disseminating any or all information related to his robot quest. Whittaker is consumed, obsessed, and disinterested in a world where robots do not or cannot exist.

His undergraduate work at Princeton, interrupted by a voluntary hitch in the Marines during the Vietnam War, helped him recognize that windows of opportunity were created by world events. A half-dozen years after Workhorse, in the wake of the *Challenger* disaster in which five astronauts died, Whittaker was lobbying NASA for robots as alternatives to human astronauts. Beginning in the late 1980s and continuing for a decade, the Robotics Institute received funding to design and build those "terrestrial test beds"—the prototypes of future technology.

The first, Ambler, twelve feet tall, had six legs which elongated independently like telescopes and were designed to navigate the boulders, deep crevices, and steep slopes NASA expected on the Moon and Mars. Then followed Dante, with eight legs, which was designed to rappel down slopes of up to ninety degrees, retrieving gas samples from boiling magma lakes in active volcanoes. Dante I died after venturing only seven meters down into Antarctica's Mount Erebus, the world's southernmost volcano, after a tether that connected the robot to its base station computers broke. Dante II successfully uploaded readings from the depths of Alaska's Mount Spurr volcano eighteen months later, making it the first terrestrial robot explorer.

William "Red" Whittaker, Fredkin Professor of Robotics at the Robotics Institute, in 2004. *Courtesy of the Red Team, Carnegie Mellon University.*

Next came Nomad, which resembled a VW Beetle, a marvelously capable robot. It could both travel long distances and be controlled from thousands of miles away to stop at a designated area and scour the landscape with a high-resolution camera. While reviewing data transmitted from Nomad's cameras during these scanning operations in the Atacama in 1997, an alert NASA geologist spotted a rock with mysterious coloration. The rock was sent from Chile to the NASA Ames Research Center, in Mountain View, California, for analysis, where it was identified as fossilized algae.

Fossils or fossilized algae represent evidence of ancient oceans, hinting at the existence of life-forms perhaps 100 million years ago. The discovery itself is rather trivial; a geologist trekking the Atacama might have easily spotted the rock and many others like it. But the fact that the scientist discovered it through Nomad vividly demonstrated that robots could become valuable tools to extend and complement a scientist's endeavors while conserving time and resources, which could prove significant for Mars exploration.

Robots had visited Mars three times before. Viking in 1976, although it was stationary, was technically a robot; it had sensors, manipulators, and other automated devices. Sojourner, in December 1996, was twenty-four inches long, nineteen inches wide, and weighed only twenty-three pounds. Unlike Viking, it moved on six-inch aluminum wheels across the Martian surface, although quite slowly; its top speed was less than one mph. Most recently, Spirit and Opportunity, NASA's Mars Exploration Rovers (MERs), which have explored Mars since January 2004, travel farther in a day than Sojourner did during its three-month lifetime, but the distance they cover is miniscule compared to Nomad's overall 200-kilometer achievement in 1997.

Nomad's last mission in January 2000 took place in the remote site of Elephant Moraine in Antarctica, where it found and classified five indigenous meteorites. But Nomad's "observation" in 1997 remained the most valuable contribution because it disproved the notion that the human scientist was irreplaceable.

Ambler, Dante, and Nomad were created in an era when ambitious NASA scientists and administrators envisioned large rovers, upwards of two tons, on Mars, exploring vast distances. Subsequently, as new administrations and reduced budgets, combined with a more realistic awareness of the challenges of Martian terrain, changed NASA's scope and direction, so too did the mission objectives change, replaced by a more economical and conservative vision for Mars. Hyperion, which means "He who follows the sun," was tested in 2000 and 2001 and represents this shift in philosophy. It weighs 345 pounds and features a concept called sun-synchronous navigation in which the rover's solar array or panel remains perfectly in line with the sun as it travels, storing energy and allowing for nearly round-the-clock operation, making for a greater science return for the dollar. But Hyperion arrived at the end of an era at the Robotics Institute and at NASA.

"Up to this point, we had been on a big roll," a faculty member told me. "In the past, before we tested one robot, there was another funded and waiting in the wings. We were on a robot roller coaster for a decade." It was a wild ride even for the city of Pittsburgh, which *The Wall Street Journal* dubbed in 1997 "Roboburgh" and put on the list of the thirteen hottest tech regions in America. Then, quite abruptly, with the delay of proposals during the change in administrations from Clinton to Bush and the terrorist attacks of September 11, funding opportunities for space-oriented ventures dried up. Red Whittaker lost his focus.

FOLKS AT THE ROBOTICS INSTITUTE had tolerated Whittaker's testy, manic periods for twenty-five years. Whittaker's mood fluctuated a half-dozen times in any given week. So no one worried—at first. This was normal Red Whittaker: Dramatic in his downs, crazed and elated when up, invariably unpredictable and relentlessly persevering. Fortitude and character had been the key to Red Whittaker's success: His ability to start something out of nothing, to follow his instincts, and, through sheer force of will, unrelenting charm, and surprising gall, to make it work was legendary. But this time, friends and colleagues noticed that Whittaker wasn't so quickly bouncing back and moving forward.

At about the same time, Whittaker's start-up company, RedZone Robotics, a commercial offshoot of Workhorse, which Whittaker had launched with much funding and fanfare a decade before, filed for bankruptcy. Whittaker's father died in 2000, which added to his depression. John Bares, a Whittaker protégé and director of the NREC (National Robotics Engineering Center), an off-campus Carnegie Mellon operating unit conceived by Whittaker to develop commercial applications for field robotics in markets such as agriculture and the military, observed of his mentor back then: "Those of us who care for him have been worried over the past few years. There have been a lot of lows for Red. He saw his company crumble, he couldn't get his proposals at NASA through; overall, he wasn't in good shape. This is a guy, a mountain climber and a Marine, who desperately needs to always seek the summit. But over the past few years he'd mostly been stuck in the Sahara."

Red Whittaker's jangled nerves and the pressure he was under to make something big happen at the Robotics Institute were evident at some of the meetings I attended in 2001–2002, as the plans for LITA (and Zoë) were taking shape. Unlike the meetings I attend in the English Department where I teach, with books and notebooks piled on tables, here at Carnegie

Mellon during meetings, the long conference table is invariably plastered with power strips, with everyone plugging in and peering at each other over laptop displays.

An OPS is anchored with a satellite communications link between the robot onsite and a science team offsite, thousands of miles away. The same basic communications procedure has been in place for MER. Spirit and Opportunity gather information and proceed to specific locations based on a plan generated by scientists on Earth. As for Nomad, scientists designed daily traverses. But if the communications link failed, or if, for any reason, Nomad was unable to function, a field team was on hand in the Atacama to rescue or repair it—an option not currently possible for MER. From the start, Whittaker advocated an aggressive home-run approach to LITA's field operations. To be legitimately Mars-relevant, Whittaker theorized, you needed to eliminate the presence of the field team during the OPS. "How else can we prove that a rover can do science autonomously," Whittaker asked, "if it's not given the chance to be autonomous?"

I was new to the Robotics Institute then and I remember that Whittaker's suggestion surprised me. Autonomy meant being independent—being able to make decisions on your own. So, of course, I thought, if the LITA robot was to be autonomous, then obviously, it should be on its own. Why even bring it up? But at the time I did not understand that although autonomy is the holy grail of robotics, it is for roboticists a muddy concept, measured in degrees. Total autonomy is a distant dream.

Thus, Whittaker's radical suggestion was greeted with a surprised silence. "After all," he added, "we are creating the first robotic scientist, are we not?" Whittaker, who is now balding and gray haired, with only a few strands of his trademark red remaining, was wearing a blue jacket with a red lining. His legs were crossed and his shoes were untied, with the laces hanging from his brown oxfords to the ground. His feet moved back and forth as he spoke.

David Wettergreen, then a thirty-seven-year-old assistant research professor, acknowledged Whittaker's suggestion by nodding—and wishing he could endorse it. How far the field team needs to be from the robot is an ongoing debate, and Wettergreen found himself consistently on both sides of the issue. He believed, on the one hand, that the field team should be at least a kilometer away from the robot during OPS to be able to prove the worth of the robot. But at the same time he realized that gathering data was the scientific mission and reason for the robot to exist, so decisions were based mostly on the best approach to fulfilling the scientific objectives. The robot they build may not be ready to do science without assistance—that was the reality to face up to.

Wettergreen is lean and soft-spoken, balding, with hair cropped a quarter-inch from his scalp, plain wire-rimmed glasses on a long, broad nose. The opposite of ostentatious, he is a walking, talking statement on behalf of the nondescript. He dresses for comfort, with boots, khakis, and knit jerseys or wrinkled button-down sport shirts with slightly frayed collars. If you see David Wettergreen in the early part of the week, he will be relatively clean-shaven, but by Friday his face is dark with stubble. I once asked if he was growing a beard. "No," he said, "I only shave on Sundays. I decided that shaving every day was a time-consuming responsibility I could eliminate from my life."

Wettergreen did both his undergrad and graduate work at Carnegie Mellon, receiving his PhD in robotics in 1995. He worked on Dante I and II. Nomad began in 1997 while Wettergreen was a postdoctoral research associate at the NASA Ames Research Center, in Mountain View, California. For Nomad, Wettergreen collaborated with NASA engineers on what he calls a "virtual dashboard" so that scientists at NASA and in other locations could see what Nomad could see. "It was designed to duplicate the experience of sitting in a car and driving." In between NASA and his return to Carnegie

David Wettergreen, associate research professor at Carnegie Mellon's Robotics Institute, in the Atacama Desert. *Courtesy of Robotics Institute, Carnegie Mellon University.*

Mellon to spearhead the Hyperion project, he was a research fellow at the Australian National University in Canberra, where he developed Kambara, an autonomous submersible robot. Kambara is an aboriginal word for "crocodile."

While Wettergreen may not be as obsessed as Whittaker, he is out of town doing robotics research almost once a week and out of the country approximately four months a year. When home, he pumps his bike up and down Pittsburgh's pot-holed hills, riding three miles from his house in the diverse ethnic neighborhood of Squirrel Hill, to campus at 8:00 A.M., and then pumping back home at 6:00 P.M. most nights to enjoy a few hours with his young children and wife, Dana. He then pedals back to campus for a few more hours of paperwork or robot brainstorming.

Both Whittaker and Wettergreen are types of critical leaders who necessarily emerge in cutting-edge institutions like Carnegie Mellon. Whittaker is a visionary type. He gets his kicks out of being somewhere first. He is charismatic and can motivate people. Whittaker's colleagues use the word "magician" when referring to him; he makes things that are absolutely impossible, possible. Whittaker believes that the world presents a series of monumental problems, which only monumental change and daring technological revolution can solve. Wettergreen, on the other hand, is the make-it-happen type. He is also visionary, but is less reliant on the power of personality. His focus is on building teams and selecting managers who can turn visions into reality, opportunity into productivity.

"One of the ways in which Red and I differ," said Wettergreen, "is that I view progress as incremental; there is an order for things." He is patient and analytical, more engaged with the notion of exploration, the impact robots can make. "In the last century, exploration meant a bunch of people on a boat who went to Antarctica, got stuck in the ice for two years and ate penguins. Half of them survived and then came back and told the story. But with robotics, that doesn't need to happen. We can explore previously impenetrable places and expand the frontier of knowledge further than anyone could have ever

imagined—and no one has to die. This is what I want to do—
expand the frontier of knowledge."

Wettergreen is not the go-for-broke guy that Whittaker is.
His career was picking up momentum, but the road upward in
the Academy could be rough. Carnegie Mellon is not anchored
by the large endowments of some of its competitors, like MIT
and Stanford. Soon Wettergreen would be seeking promotion
and tenure, and he needed a strong track record to stay viable. If
LITA was successful, it would be a feather in his cap—a high-
profile entry in his already impressive CV. But in some very significant ways, Carnegie Mellon robots are different from the
robots developed by Jet Propulsion Laboratory (JPL) or the
Johnson Space Center in Houston projects, which can exceed $1
billion. "We do developmental work," Wettergreen explained.
"We determine what is possible—we test future concepts."
Most of the robots developed at Carnegie Mellon are for the
possibility of tomorrow rather than the reality of today.

University administrators—and Wettergreen—had higher
aspirations for the Robotics Institute, envisioning those $1 bil-
lion grants and the prestige that comes with it. To this point, the
closest Carnegie Mellon projects had gotten to space was the
autonomy software being used on the MER, which came from
an algorithm developed at Carnegie Mellon by former student
Mark Maimone. He was one of the many dozens of Carnegie
Mellon alumni at NASA contributing to the space program.

That was fine, but Wettergreen wanted Carnegie Mellon
and its software more directly involved in the action. He
hoped that the Atacama project could help trigger such a leap,
and that Carnegie Mellon could be a new and vital force in
determining how planetary exploration would be conducted
beyond the next decade. The MER robots were seeking evi-

dence of water on Mars—a significant leap forward—but to this moment, no robot was sophisticated enough to discover signs of life.

The Field Robotics Center, Whittaker's wing of the Robotics Institute, had recently partnered with two aerospace companies and submitted proposals to NASA for two major flight projects that essentially would have elevated Carnegie Mellon to the status of JPL. They were rejected because "NASA didn't feel we had the experience to take on such ambitious projects." By "experience," Wettergreen was talking about adults— veteran programmers and administrators—not the grad students or junior staffers currently writing software and designing and creating software and hardware. Yet the very strength and brilliance of the Robotics Institute was in its young, naïve population. Carnegie Mellon is identified as a breeding ground for off-the-wall ideas and concepts that mature programmers and engineers might never attempt or commit themselves to. But at the same time, the spontaneity of the place limited opportunities.

"We are a pretty transparent organization," Wettergreen told me. "People can come and visit us and see the chaos happening day and night. Mistakes are made. Corners are cut. They will also see a lot of excitement and energy and sometimes, incredible innovation. Having been one of those children, it amazes me what can be accomplished. There is something about this place that allows young and inexperienced people to ramp up fast."

But turning young people loose in a multimillion-dollar laboratory requires careful oversight and regular down-to-earth direction. "When I work with senior staff I only have to say 'we are going in this direction' and immediately they will know what they need to worry about and what they don't need to worry about—and what to do first. For the junior people, you have to define the challenge of the task on a regular basis, explaining what will be difficult and what will be easy, and you have to watch them closely so they don't go off into wild-eyed irrelevant directions. Of course," he added, "young or old, to be successful, you have to basically commit your life to the work." Whittaker was the perfect model of unwavering commitment.

Going for the home run and proposing a total Mars-relevant effort as Whittaker was advocating would be daring and dramatic and true to his mentor's philosophy and the Carnegie Mellon approach, Wettergreen acknowledged, but dangerous and unnecessary. The NASA evaluators reading the proposal would understand what a long shot total autonomy with a rover would be at this point—and probably reject it. Besides, to be practical and true to the mission, ASTEP demanded scientific results. While robots are an important element, they will be, in this context, created to do science, not to prove that autonomy is technologically possible. For ASTEP, science is the end game—not robotics.

Ethnocentrism has been an ongoing problem among roboticists—zealots who cannot see beyond the challenge of the technology. For the most part, robots do not or should not exist to prove themselves. Rather, they must exist to support useful endeavors, such as ASTEP. The rule of thumb cannot be robots for the sake of robotics, but robots for the sake of science.

The new and enlightened roboticist is neither code monkey nor wrench jockey. They are humanists and inter-disciplinarians—deep thinkers. Robotics is a composite of computer programming and cutting-edge engineering, but together, moving from problem to solution, it is the ultimate intellectual exercise in technological creativity—a mix of substance and outlandish ideas and improvisation, similar to jazz. Bottled up on a daily basis in this hypertechnological pressure cooker at Carnegie Mellon, Wettergreen found all of this intellectual activity inspiring, compelling, and addictive. He was more than just a robotics engineer; he cared deeply about the scientific breakthroughs to be achieved in the upcoming robotics age.

Throughout the meeting, Whittaker continued to lobby for total autonomy, while Wettergreen was consistently resistant. "If something happened to the rover during the traverse," said Wettergreen, "we absolutely must have a person in the field around to deal with it. I don't want to set up our team so easily for failure."

Whittaker's next comment was indicative of his personality, his history, and perhaps his frustration over being stalled at the peak of his career. By going for broke, he countered, "We are setting ourselves up for success. Let's be true to our mission: To make the first autonomous discovery of life with a robot. No one," he added, "has done this before." But the answer to Whittaker at that moment went unspoken: It echoed through the silent emptiness of the room.

Red Whittaker may have lost his go-for-broke pitch to the more practical members of the LITA team, but the irrepressible mountain climber and ex-Marine was already soaring skyward chasing another idea that most veteran roboticists considered impossible.

CHAPTER THREE

Groundhog

IN JULY 2002, THE WORLD WAS TRANSFIXED ON THE tiny town of Somerset, Pennsylvania, as rescuers drilled into nearly 300 feet of dirt and rock in a desperate attempt to save nine miners trapped underground in the flooded Quecreek mine. For a while, rescuers didn't know if the miners were alive until they heard tapping from survivors—signals coming from what turned out to be a small air pocket. After seventy-eight hours of ceaseless digging and pumping, all nine men were rescued. Their stories have been told repeatedly and captured in a book and a made-for-TV movie. Their plight and bravery affected anyone who followed the rescue, most especially Red Whittaker.

Somerset is not far from Red Whittaker's farm, and perhaps sixty miles from the Carnegie Mellon campus in Pittsburgh. Thinking about the near tragedy soon afterward, Whittaker had a sudden inspiration. He began making phone calls to friends in the mining industry, asking questions. Whittaker learned that Quecreek had been flooded because the miners, guided by old, inaccurate maps, had excavated too close to the adjacent long-flooded Saxman Mine, causing the walls between the two mines to collapse. This is a serious problem in the mining industry which no one had addressed—until now. This is also, quite typically, the way in which Whittaker dives into a project: No money, no plans, no staff, no permission from federal or state agencies to go into the mines he is intending to map. Consumed by his vision and the fire-in-the-belly to think he can make something out of nothing, Whittaker turned to the only means of support remaining—what he terms his "children's crusade" or "rookie" option.

He immediately designed a course for students and made mine mapping with a robot the course project. The students he attracted, an even mix of engineering and computer science grad students, were talented and, after listening to Whittaker's idea, motivated. After all, that's why they came to Carnegie Mellon. This was the school's reputation: To enable students to get involved in the heat of the robotics action, in the field and the trenches, and to work with and perhaps be mentored by the famed father of field robotics. These young people are the bone and sinew of the institution and the reason the robots and the Robotics Institute exist.

Zachary Omohundro immediately bought into the concept of what Whittaker called "technological swashbuckling in coalmines." Of Scotch-Irish descent, he is tall, slender, and dark, with eyes that seem constantly in motion. "I applied here and at MIT and came to visit both campuses. At MIT, they gave me a PowerPoint presentation about theoretical frameworks for future systems—and that pretty much made my decision. At Carnegie Mellon there were robots everywhere."

"Robots everywhere" is an accurate description. The High Bay on the ground floor of Newell-Simon Hall, headquarters of the Robotics Institute, resembles a small aircraft hangar. That's where most of the large robots are actually put together and tested—beasts like Hyperion and Nomad, all sitting like warriors resting their thunder, patiently waiting to get back into action.

At any given moment, there will be robots here from many research groups, including the Field Robotics Center, the Vision and Mobile Robots Lab, the Robot Manipulation Lab, the Helicopter Lab, the Advanced Mechatronics Lab: Zooming robot Segways learning to play soccer; robots on treasure hunts; Nursebots that care for patients in hospitals; robots who look like Lara Croft (of *Tomb Raider* fame); robots who attend meetings and give speeches. Robots named Pearl, Houdini, Enviroblimp, Bullwinkle, Grace, Gyrover, Ferret, Demeter, Rhex, Xavier; robots that build other robots; robots that do origami; robots destined for museums as tour guides; robots that squirm along the ground like snakes; robots made from golf carts; robots that creep like caterpillars; robots that are pogo sticks; robots that roll, jump, crawl; robots that resemble beachballs; robots that look like eyeballs; robots that fall apart when they move; miniature robots called milibots—all zipping back and forth with little regard for pedestrians. Collisions are known to happen. Civilians venture into the High Bay at their own risk.

Nearly all of these zipping and zooming little creatures are being controlled by disheveled young men sitting on the floor, on folding chairs, tool cases, portable coolers, huddled over laptop computers, peering at displays with a squinting, manic intensity.

Take Chris Baker, a twenty-four-year-old native Pittsburgher, who did his undergraduate work at Carnegie Mellon. Baker is now working on a master's degree and will soon enter the PhD program in robotics. His father is a general contractor for whom he had worked every summer, so he had certain basic skills—a knowledge of wiring and welding—that the average graduate student lacked. He and Omohundro, who grew up in Minnesota and gained experience building a robot during his senior year at Rice University, were what, according to Baker, Whittaker called "fresh blood": Young people willing to work endlessly under difficult circumstances, which was the gauntlet Red Whittaker put his students through.

Baker, Omohondro, and around a half-dozen other students knew that mine mapping with a robot might be the opportunity of a lifetime, although the reality of the experience was, at the very least, sobering. It began with a marathon sleepless weekend of work, starting with a two-hour drive out of the city to a golf cart graveyard.

"It was just like a car junkyard," said Baker. "There were golf carts in many states of disrepair. We selected a golf cart, threw it in the back of a truck, took it to Red's farm, and tore it apart." They worked nonstop, alongside Whittaker, over the weekend. "We declared success," said Baker, "when we hooked it up to the batteries and the golf-cart-now-robot moved." They named their new mine-mapping robot Groundhog. "Then we could go home and go to bed."

This was only the beginning of the task of turning a golf cart into a working groundhog; weeks of designing and redesigning, writing and debugging software, tearing Groundhog back down to its bare frame to strengthen and rebuild it—thousands of tedious, sleepless hours. Red Whittaker, like George Patton in World War II, ignited among his troop of rookies a nearly maniacal dedication, as well as an overpowering exhaustion.

I have had many conversations with Whittaker. When we talk one-on-one, he sometimes rambles, stumbling over his words and going off on tangents that seem pointless. But groups of young people magically transform Whittaker. A dynamic articulation suddenly emerges. At a class I attended during the time Groundhog was being tested, Whittaker attempted to convince his weary, overworked students to write a proposal for a Groundhog-related project to take place after Groundhog had been tested.

The students, clustered around a long rectangular conference table, listened quietly, hardly moving, as Whittaker described the project and all of its potential benefits. It was difficult to know if they were enthralled by his rhetoric or stunned by his presumption. But they were clearly in his grasp.

"Nobody around this table looks particularly bored with life and needs something more to do," Whittaker commented, pausing, sweeping his piercing gaze around the table, peering intensely into each and every eye. (Describing this look of deep scrutiny, one student said it is "like Red is grinding himself into your soul.") "You don't have to do it," Whittaker continued, now staring out the window into the distance, seemingly visualizing the future. But then he leaned forward, wrinkled his brow, and raised his forefinger in warning: "Oh, by the way. The opportunity won't come again. This is your moment." He paused dramatically. "What are you going to do with it?"

This is the Whittaker style and approach, building up a project and then challenging his mostly male students (all male in this case), counting on their testosterone and their competitive, overachieving personalities to take the bait. After discussing the project for a few more minutes, Whittaker toned down his rhetoric, talking quietly, almost soothingly. Then he lapsed into a long silence and stared dreamily into space.

"Red's way," said Chris Baker, "is to say, 'We need somebody to do something,' and then there will be this really intense quiet until somebody says something to break the silence." Whittaker was usually willing to outwait the discomfort he had caused.

Whittaker, according to Omohundro, has an "optimistic zone" around him. "Once you get into it, it is kind of hard to disagree with him." He is "inspiring," said Omohundro, "and terrifying."

Baker and Omohundro, along with two or three other students, built Groundhog from scratch. "Red has an obsession that both Chris and I understand," said Baker. "People who understand tend to gravitate toward him while people who don't tend to drift very far away."

Indeed, many of the graduate students and PhD candidates I talked with have come to Carnegie Mellon at least in part because of Whittaker's legendary status. But once on campus, they can be put off by his unrelenting expectations. His aura is magnetic, but his presence is frequently avoided, either because his moods are unpredictable—he can lash out unexpectedly—or, worse, he can charm or manipulate you without your ability to resist.

"Meetings with Red don't entail conversation," a Carnegie Mellon graduate engineer who has worked with Whittaker told me. "They are more like motivation lectures. Anytime I talk with him, it feels like he is trying to figure out how to get the best results from me. So if criticizing—hammering on you—gets the results, or telling you that you are great is going to get the best results, then that is what he will do."

But Whittaker gets a tremendous amount out of the people who choose to remain with him, despite his tantrums and manipulations. Omohundro, Baker, and many others who are swept up by a Red Whittaker project talk in terms of "thirty- and forty-hour days." They mark significant moments in the project by saying, "That's when we could go home and sleep."

Neither Baker nor Omohundro were unaware of the significance of their hard work and the purpose behind it: "Sometimes we forget the fact that no one has ever put a robot in a mine before. We are students and we are making history," said Baker. "That's awesome!"

Making a robot operate autonomously in a mine presents special challenges, perhaps even more perplexing than operating one in outer space. Robots are more effective on level ground, in an open environment with good uniform lighting. GPS (global positioning systems) or satellite imagery will help position and guide a robot—even in the Atacama or on Mars. But an unmapped mine is a dark, unresponsive mystery.

"The robot sensors are less reliable underground," Aaron Morris explained. Morris is short, solidly built, and strikingly handsome with a pale, baby-faced complexion. His words are laced with the flat, guitarlike West Virginia twang where he grew up, although, ironically, he had not ever entered a mine before the "subterranean" robotics course. But he had learned a lot about his home state since entering Carnegie Mellon and the integral relationship robots might have with the mining industry.

Since it was an exploration task—the mine into which the robot entered is a black box—there was no way to plot where the robot was going or anticipate what it would encounter: fallen timbers, flooded shafts, abandoned machinery. Morris explained, "You program a robot to think through a problem and make intelligent, accurate decisions. Mistakes are often irreversible. A robot lost in a mine can usually not be rescued and may never return."

Groundhog sees by building a two-dimensional map of where it is going, called a "cost map," meaning that, as it travels, the code allows it to see the obstacles in its path and attempt to avoid them or analyze the potential cost of going forward and over or under them. Robots will not recognize an object it sees. A harmless cable hanging from the ceiling and a dangerous pillar may look roughly the same to Groundhog.

The word "see" doesn't accurately capture the reality of the robotic vision and navigation system, by the way. Groundhog will find its way into and, if all goes well, out of the mine by counting and memorizing intersections. "That's what you, anybody, would do if you were in a cave. You come to an intersection, and you are immediately at a decision point—left or right or straight. If you go straight, you will leave a marker and therefore when you return you know you have been there before."

I asked Morris about the maps I had seen Groundhog and other robots make—vivid, colorful illustrations of the inside of mines or other areas it or other robots may have traversed. "We build a 3-D model after the fact for easier viewing, using the 2-D data the robot has gathered." The maps, he stressed, are for people to view—not what a robot will see.

Veteran roboticists at Carnegie Mellon, like Whittaker and his protégé Wettergreen, insist that they do not feel any connection to the robots or the code they conceive and create. But rookies like Morris can't identify with the ambivalence of their teachers.

"Red likes to say that humans don't have a personal connection to the robot—but he is wrong—there is always a personal connection to the robot. Code is a reflection of the person who writes the program, so how could I not feel connected? These robots are extensions of us. If they don't do what we program them to do, we are going to feel a great sense of failure."

"Are you thinking for robots when you program—write code—for them, or are you helping them think?" I asked. This is a key aspect of robotics, and a subject of constant debate and reflection. Are the robots simply reflecting the programmer's wishes—or, when they traverse a mine (or Mars), are they acting on their own?

Morris smiled and shook his head slowly back and forth, weighing his response. "This is something I ponder all of the time, and I don't really have an answer. 'Thinking,' 'intelligence,' 'cognition' are such vague words. Programmers writing code lay down a set of rules for a robot to follow, so maybe the robots are simply following orders." But then he paused to reflect and shrugged. "But who's to say that thinking, for all of us, is not following a set of rules? People are consistently determining and breaking boundaries."

But "thinking" or following rules like people may not always be the objective, many roboticists contend. Robots have sensors, global positioning systems, gyroscopes; they often know a lot more than people about a particular situation or problem and can jump start the cognitive process. To maximize efficiency, humans and robots may very well need to think differently.

Morris did his undergraduate work at West Virginia Institute of Technology, an independent school recently absorbed by West Virginia University. He started out in engineering because "[his] father thought it was a good idea." But he moved to robotics as a graduate student at Carnegie Mellon, and now after five years of study, he is working on his PhD. He does not consider himself a typical robot-rookie, however. "After five years, I have become an old man here." He is twenty-six years old.

But, like his fellow students in this conference room listening to Red Whittaker attempt to motivate them, he couldn't seem to focus on much of anything else. "We eat, sleep, and breathe robotics," he said. "I wander into a convenience store and think of all of the ways robotics will be important. We become obsessed," he said, waving his arm like a flag. "We aspire to change the world." This seemed to be a quality—or a malady—that most of the rookies not only under Whittaker's spell but also throughout the Robotics Institute shared. Which is why Morris was the first to raise his hand in the classroom that afternoon when confronted by Whittaker's new proposal. "We have to do it," Morris said.

Whittaker smiled. "That's one voice. Anybody else?"

Four more hands were raised. "Five of nine, that's terrific," he proclaimed.

Then Whittaker hunched over and peered intensely into space, foreseeing the future and sealing the deal by anointing his players. "I actually pity anyone who is competing against you because . . . who could run with you? Who could touch you?" He went on like this for a while, then asked: "How do you get thirty good pages?" The proposal must be thirty pages long.

"Lots of pictures," one student yelled out.

Whittaker laughed: "How do you get thirty great pages?"

"Write the first one," another student replied, quoting his mentor.

"OK, is there anything else to say?" Whittaker waited. No one answered. "Going once, going twice . . . Who is leading?"

"I guess I am," a staff member mumbled.

"If you have a sense of conviction, you don't have to guess," Whittaker replied.

Whittaker waited again, but the specter of yet another exhausting, never-ending commitment was now, suddenly, upon the "new generation," and they were, perhaps, momentarily paralyzed. "OK, make it work," Whittaker declared, as he jumped from his chair and dashed out of the room.

Later, when I asked Aaron Morris how he felt about being confronted with yet another daunting job to do, he shook his head, spread his arms as if feigning helplessness, and laughed: "We raise our hands. We follow. We take that gauntlet. We like to do what people say is not possible."

As it turned out, the proposal they committed to write on that day was written and made the deadline—and then was rejected. But according to Aaron Morris, "The hard work, even though it was completely over the top, kept us focused."

CHAPTER FOUR

Bummed and Elated

THE MATHIES MINE HAD BEEN FILLING WITH WATER, poisoned with the residue of the chemicals used for mining, ever since it had been sealed off in 2002. Most of the water drained from the mine was treated at a nearby plant on the Monongahela River, but the water flow had been increasing, and the state Department of Environmental Protection (DEP) was assessing the need to run a pipe through the mine to remove excess drainage. The possibility of an accurate map to help decide how to place the pipe eventually led to the approval of Carnegie Mellon's request to enter Mathies with a robot.

The Mathies Mine was unique in that there were two entry portals, 3,500 feet apart. If things went smoothly, Groundhog, which had grown from a golf cart to a squat, 1,600-pound ATV look-alike, would creep into one portal and exit the other in approximately four hours—having made a map of what it saw inside. Anything could happen to Groundhog in the interim, which was why DEP and Mine Safety and Health Administration (MSHA) officials warned Whittaker, and his colleagues and co-teachers Scott Thayer and Sebastian Thrun, that if Groundhog did not come out of the mine on its own, then it would be lost forever. No one from Carnegie Mellon— no one unauthorized—would be permitted to rescue it. This seemed to be exactly what Whittaker wanted to hear in order to heighten the stakes for his rookie roboticists. "This has to work perfectly," he announced each time he made an appearance in the High Bay or the machine shop. "Or we will lose Groundhog to the mine."

To Whittaker, who had fathered more than sixty robots during his thirty-year career, losing Groundhog would have been a technological setback but not a major loss. "I don't get happy about robots or feel sorry for robots. They are not like little old ladies or puppies. They are just machines," he said. He once advised a group of students and faculty not to waste feelings of love on a machine. "They certainly don't have the same feelings for you."

But the young engineers and programmers who had created Groundhog from scratch had much more at stake than did their ambivalent mentor, Red Whittaker. The specter of losing their first "real" potentially autonomous creature took on catastrophic proportions. As "Groundhog Day," May 30, 2003, neared, the students found themselves working harder than they had ever imagined possible. Sleeping even two hours a night was a luxury many of the nine students, applying the final touches to Groundhog's body and debugging the programs, could not afford.

They realized the night before the Mathies operation that the software had been rewritten and debugged so often it had never been tested completely on the robot. And that night, when they finally attempted a test, Groundhog suddenly stopped dead. Every time it was rebooted, Groundhog would stop again. "Finding what was tripping it off was like finding a needle in the haystack," said Aaron Morris. At 2:00 A.M., Groundhog was in pieces, spread out all over the floor of the High Bay.

Aaron Morris told me later that because of the mess and confusion in the High Bay most of Groundhog's testing took place in the fancy atrium area two floors above the High Bay, in the entrance to Newell Simon Hall. During the day, bright sunlight, a rarity in normally gray Pittsburgh, will often cascade down from skylights and through a wall of plate glass into the atrium. There is a cafeteria, featuring tasty, economical Asian cuisine. Pizza and deli food are also available. But at night, the atrium is abandoned, except when invaded by desperate, sleep-deprived roboticists. "We moved all of the furniture out of the way when everyone left the building, including the tables from the cafeteria. We ran Groundhog until the early morning, then put the tables and chairs back where they belonged before sunrise." It was, as Whittaker might have described it, a total Stealth operation.

Eventually that night before Groundhog Day they discovered the answer: the potentiometer, the sensor that communicates the angle of the wheels, had broken. They could not repair it in time or find a replacement, so they had to cannibalize a potentiometer from another robot. By the time Groundhog was put together and guided up the ramp and into the rented U-Haul box truck for the drive to the mine, it was 5:00 A.M. Everyone was bleary eyed but pumped up with anticipation—and trepidation.

As I watched the drama of Groundhog Day unfold, I realized that this scenario of last-minute catastrophe and emergency surgery was a pattern roboticists at Carnegie Mellon could not seem to avoid. This was the chaotic atmosphere David Wettergreen had described in explaining why NASA might hesitate to give major missions to Carnegie Mellon. Yes, the job at hand was getting done, no doubt about that. And the determination and creativity the students had displayed in testing and preparing Groundhog in the middle of the night in the atrium were remarkable. But $1 billion Mars missions cannot be run on a seat-of-the-pants process. "If we're ever so fortunate to get one of those big grants," Wettergreen had admitted, motioning toward the High Bay, "I can assure you I wouldn't do the work here."

Whittaker, who is usually willing to work as hard and as long as any of his most dedicated students—and more—had not been on the scene very often in the days leading up to Groundhog Day, although Thrun and Thayer were putting in nearly as much time as their students. Still, Whittaker's indefatigable presence, no matter where he was physically, was a factor. He is, when he wants to be, the consummate communicator.

Most every Whittaker project seems to include a mission time-line beginning with a significant number, such as 100 days to Groundhog Day, etc. In this manner, he can heighten the suspense and gradually count down and increase the pressure on his staff and students to work harder. For most projects, he routinely establishes an e-mail log to provide project news and impart eloquent observations from his own experiences and from inspiring quotations by mostly famous people, such as:

> *Great things are not done by impulse, but by a series of small things brought together* —VINCENT VAN GOGH

...

*The longer I live, the more I am certain that the great differ-
ence between the great and the insignificant, is energy—
invincible determination—a purpose once fixed, and then death
or victory* —SIR THOMAS FOWELL BUXTON

...

*It is not the critic who counts, not the man who points out
how the strong man stumbled or where the doer of deeds could
have done better. The credit belongs to the man who is actually in
the arena; whose face is marred by dust and sweat and blood;
who strives valiantly; who errs and comes short again and again;
who knows the great enthusiasms, the great devotions, and
spends himself in a worthy cause; who, at the best, knows in the
end the triumph of high achievement; and who, at the worst, if
he fails, at least fails while daring greatly, so that his place shall
never be with those cold and timid souls who know neither victory nor defeat* —THEODORE ROOSEVELT

This last quotation was sent to all parties interested in the
mine-mapping project in defense of a colleague who failed to
get Groundhog access to the Saxman and Quecreek Mines,
their initial objectives. Whittaker is consistently loyal to the
people who work for him, but he has no tolerance for quitters
or for people who are not highly motivated. He pays rigorous
and sometimes fawning attention to everyone who might
impact his project in a positive way. He cultivates corporate
executives because of the resources at their disposal. And he
panders to the media, which have a totally different but
equally essential way of affecting the success of a robotics
project and highlighting his own contributions.

Watching Whittaker perform, it is difficult to say when or
if he is being devious and manipulative—or just his eccentric

and temperamental self. As far as I could tell, he is neither evil nor saintly. He simply wants to achieve, and he will do or say whatever is necessary to make his mission a success—*or make it seem like a success*, if it doesn't quite fulfill his very expansive expectations and predictions.

HERE WAS THE MOMENT Groundhog began its traverse into the Mathies Mine portal.

The students stood stiffly inside an imbedded railroad bed, which led into the mine. Surrounded by grass and wildflowers, they resembled toy soldiers in their red and yellow hard hats and goggles and their black Groundhog Day T-shirt uniforms. Instead of rifles or lances, they were clutching their beloved laptops at their sides.

For a while, we were silent, listening to the whine of Groundhog's laser sensor—which scanned the walls and ceilings of the mine, took a two-dimensional image, and determined whether it was safe to move—and the electric motor which powered the vehicle forward. In addition to visualizing the path ahead, Groundhog would also have to decide, based on the data it scanned and collected on its cost map, whether it was best to climb over a small obstacle in its path or go around it. A large obstacle may force retreat. This jerky, stop-and-go process of thinking—scanning, evaluating, and moving—would happen hundreds of times over the next few hours, as long as Groundhog, with a battery life of nearly six hours, stayed alive.

"Feeling a little anxious?" Aaron Morris was asked as he stared at the slowly retreating robot.

"A little bit."

"Get a little sleep last night?"

"Of course not."

"You say that with a smile."

"That's all you can do at this point."

More silence, as the shadow of the low ceiling of the mine began to envelope the robot, water dripping from the ceiling onto its hood. Groundhog's tires were caked with mud.

"We might be looking at Groundhog for the last two seconds," Whittaker said matter-of-factly. He was wearing a brown sweatshirt and a battered Australian campaign hat. Then he added: "It might come out the other side or it might encounter conditions that will trap it inside. No one knows."

The students began to back away from their baby; it slowly stopped, scanned, jerk-started, and whined as it was swallowed up in the darkness.

"Kiss it good-bye," Dave Ferguson said. Ferguson had helped program Groundhog. He had come to Pittsburgh to do his graduate work from his native New Zealand.

"We put in a lot of time and not a lot of sleep," said Chris Baker.

"If it doesn't come out, then does that have any bearing on your grade?" Aaron was asked.

"I think we get an incomplete," he joked.

"Can you still see Groundhog?"

"I can see it."

"Looking happy?"

"Absolutely."

Inside the back of the now empty U-Haul, Aaron, Dave, Chris, and Zackary have set up an operations headquarters to monitor Groundhog's progress. They all gathered around the TV monitor. This was the cardiogram—Groundhog's heartbeat—which registered a bleep and a blip every two seconds on the display. Groundhog's blipping and bleeping will comfort them.

"I feel just like how I will feel when my kid is going to college—excited but sad at the same time," Scott Thayer said. Thayer, a native of Tennessee, is heavyset, bearded, and scruffy, a mirror image of all the students he has helped lead. Thayer had worked night and day alongside the students. "You guys are fucking awesome," Thayer said.

"That," he was told by Scott Baker, "remains to be seen."

The students laughed, joked, and worried, while they stared at the monitor, hardly moving for the first half hour Groundhog was gone.

"If Groundhog drank beer, what kind of beer would it drink?"

"Not IC Light," someone replied, referring to the local (Iron City) brew.

"We have to stop watching this monitor. It is making us crazy," Thayer said.

"I will be equally crazy," said Zackary Omohundro, "if I stop watching."

But soon there was nothing to watch. The blip disappeared. The display turned as black and blank as the dark hole to the entrance of the mine into which Groundhog had disappeared. Everyone had expected Groundhog to venture so far into the mine that its signal would eventually fade away, temporarily, until it either turned around and retraced its steps or traveled past the halfway point and began heading to the opposite portal, but this was too soon for that to happen.

And then, as they sat there, their eyes glued to the blackness of the display, glancing at their wristwatches and staring anxiously at one another, they realized that now *too much time* had gone by, waiting, without the signal's reappearance. Something was wrong, terribly wrong. Groundhog was missing. Absent without leave.

Some of the students remained in the truck attempting to make contact with Groundhog's wireless signal with their laptops, while collecting the information Groundhog had so far relayed back to them. Others jumped into a car and drove a mile down the road to the operations tent at the other mine portal, where they had expected Groundhog to reemerge. There they found Whittaker and Sebastian Thrun, a slender, balding fellow with a baseball cap and a Patagonia jacket, analyzing the data so far transmitted by Groundhog. The tent had been set up by the Carnegie Mellon public relations staffers. The local newspapers and TV stations were on hand. In Pittsburgh, Red Whittaker and Carnegie Mellon robotics are always news.

But Thrun led the way at first. He stood up to announce in his thick German accent "the good news and the bad news," which seems, in the emerging world of robotics, par for the course. Nothing ever works right, unless something else goes wrong. Nothing works the first time and very often nothing works the first hundred times. And if exceptions occur, if something goes right immediately, it will probably fail when the experiment is repeated. The balancing act—the good and the bad (and sometimes the ugly)—was nearly endless.

Thrun told the group that the data so far transmitted demonstrated that Groundhog had traveled 308 meters and that it had "navigated wonderfully." That was the good news. The bad news was that Groundhog was lost. "We don't have communication with the vehicle anymore."

Thrun explained that Groundhog had encountered a fallen beam 178 meters from the entrance to the mine. Using his laptop display, he showed the video of the beam, transmitted by Groundhog, about a foot high, lying across the mine floor,

Groundhog and the research team at Mathies Mine near Elizabeth, Pennsylvania, May 2003. *Courtesy of Carnegie Mellon University.*

from wall to wall, making it impossible for Groundhog to progress. Disappointment? Yes. But here was a small but astoundingly clear image emanating from his display—a wonderful grid—a detailed three-dimensional map of where Groundhog had been—evidence of the potential of using robotics to map mines or for search-and-rescue operations. Robots could certainly go where man could not and provide eyes and insight into where man was unable to see, Thrun stated. Thrun was smiling and Whittaker was nodding.

Unable to find a way around the beam, Thrun continued, Groundhog had turned and retraced its steps 130 meters back

Groundhog emerges from an abandoned coal mine.   *Courtesy of Carnegie Mellon University.*

toward the entrance. And then, mysteriously, suddenly, it stopped—dead. "It is stuck, and we don't know why." The communication link between Groundhog and its creators had died. One of the reporters then asked if the experiment should be considered a failure. Before Thrun could answer, Whittaker, who had been sitting quietly behind Thrun, replied that the robot had made the appropriate decision when it encountered the obstacle and turned around and went back.

"So was it successful?"

"It did the right thing—it found an obstacle it could not squeeze by, and then turned around and retreated." This evidently was the way in which Whittaker would spin the experiment to make a tepid result seem triumphant—a process he had perfected. Misleading, perhaps, but considering his reliance on other people's generosity and support, vitally necessary.

"Well, it is gone now," Dave Ferguson observed a few minutes later as they returned to the U-Haul to begin breaking camp and cleaning up. He sat down, yawned, put his head back, and closed his eyes. In the back of the truck, Aaron Morris was leaning against the wall, staring off into space. A few of the others were also hanging around the truck; the disappointment in their faces was sobering.

"Don't look so unhappy, guys," said Thayer. "You did a great thing today. You got such a wonderful map. We have proved our point."

"And we sacrificed a robot," said Zackary Omohundro.

An hour after his students had hit the bottom of the depression barrel, Whittaker persuaded an official from MSHA to break regulations and venture into the mine to see if Groundhog had survived. The MSHA official found Groundhog within ten minutes, flipped a switch, rebooted Groundhog, and escorted it back through the mine and out the entrance.

The explanation regarding what had happened to Groundhog in the mine was simple. Typically, the problems roboticists confront with new robots are rather elementary. Groundhog's computer crashed. Avoiding such mistakes or learning to deal with them on the spot were big issues, difficult to predict and nearly impossible to solve without direct access to the robot. Baker and Ferguson devoted three months to revising and debugging the software. In subsequent tests through 2004 and 2005, Groundhog reentered Mathies and made it from one end of the mine to the other—repeatedly.

The Mathies event was exceedingly exciting, and the students had been happy to have Whittaker and Thrun spin the event into a consummate technological victory. But the unrelenting pressure on the rookies of the new generation had been overwhelming, good perhaps for character development, but hard on young nerves. As Scott Baker so succinctly explained the up-and-down rhythm that characterized his day: "Bummed, elated. Bummed, elated. I can't take it anymore." But Baker, Ferguson, Aaron Morris, and the entire Groundhog crew gritted their teeth, steeled their nerves, and followed their leaders home to the High Bay for beer, pizza, and celebration.

Whittaker, the true believer and cheerleader, captured their frustration and their relief in his characteristically inflated and flamboyant language: "With all their purpose, power, and ferocity, this generation exhibits grace and honor to one another before their time."

CHAPTER FIVE

A Lack of Vision

I<small>T IS 6:45 A.M., THE MORNING AFTER OUR ARRIVAL IN</small> the desert, and I wake up to the sound of Mike Wagner, singing the theme song to *Mister Rogers' Neighborhood*. Wagner is twenty-seven, small, solid, and balding, with a perpetual, wide smile that makes him seem more like Howdy Doody than a Fred Rogers aficionado.

It isn't surprising that Wagner knows all the words to the song, from its beginning, "It's a beautiful day in the neighborhood," to the last inviting stanza, "Won't you be my neighbor?" A senior research associate at the Robotics Institiute, he is one of the youngest old men or the oldest young men I have ever met. While Wettergreen, the official leader, is quiet and to a certain extent aloof, Wagner is the glue of the project. Friendly, frequently teasing and joking, always congenial and helpful, Wagner can be intense, inquisitive, and combative. At the same time, he is a first-rate team player, especially here in the field. I have watched as he has matured with the project, taking on more of a leadership position and often forcing decisions—or making them on his own—when no one else is willing to commit. He often defuses tension with his jokes or his willingness to be silly and sing a song, as in this instance.

Most of the roboticists at the Institute are accustomed to working on their own—a necessity in a situation in which most aspects of the robot depend on the development of various software or hardware components. Wagner works toward the integration of the overall team. "I am not into one particular element of the robot," he says. "I am into bringing ideas and concepts together and making the whole thing work." He is not necessarily into robots, either—or space, as are many of his other colleagues. "I want to invent things, and have new experiences."

Wagner is a veteran of three other field experiences. As a Carnegie Mellon undergraduate in 2000, he accompanied Nomad to Elephant Moraine in the Antarctic. Less than a year later, Wagner was in the Arctic on Devon Island with Hyperion, testing the concept of sun-synchronous navigation. In 2003, he accompanied Hyperion to the Atacama, on the LITA initial shakedown tests. Hyperion was used in 2003 as a software testbed for Zoë in 2004. Wagner's journey from robot to robot and project to project at such an early stage in his career leads to a nonchalant ambivalence not unusual among these young roboticists. "It is always fun to build a robot," he told me when we first met. "The coolest thing is to watch it drive around on its own and know that your brain and effort has had a small part in making the robot come alive."

This morning, however, after his song, he is contemplative, as he brushes his teeth and watches the emerging dawn through a thick blanket of white fog. A few people have not yet crawled out of their tents, but most everyone is slowly congregating around a stack of five-gallon jugs of drinking water. Although rudimentary showers are available at the salt mine where we eat breakfast and dinner, most of us keep clean with Dry Wipes. Last year, LITA team members competed to see who could last longest without showering. Wagner won the no-bathing battle, a silly contest, he admitted. This year, since arriving two weeks ago, water has yet to touch his body.

At this point in the Atacama, however, Wagner is not having very much fun. His early-morning nonchalance is gradually replaced by a dead-serious intensity. Fixing Zoë's vision system is his responsibility, and although he is clueless at the moment as to the source and the solution to why Zoë is not seeing what it clearly is looking at, he is enough of a veteran to know that it is too soon to worry. His friend and colleague Stuart Heys, the engineer responsible for designing and building Zoë and who shares a tent with Wagner, on the other hand, is feeling increasingly burdened by the pressure of the upcoming OPS.

Stuart Heys grew up in the Philadelphia suburbs, a frail-looking splinter of a boy with a reedy voice and a nervous, self-deprecating manner. He entered Carnegie Mellon, like his father, with the intention of majoring in chemical engineering. After his first term, he transferred into material science. "But," he admitted, "I hated them both." He finally settled on mechanical engineering. Not a choice, but a process of elimination. Carnegie Mellon's robotics program is the finest in the world (MIT debates this) and its school of computer science is among the top four (along with MIT, Stanford, and Caltech), while the engineering program is nearly as prominent and is especially popular because it serves as a bridge to computer science and the applications-oriented work done in robotics.

Heys did not have robotics in mind, however. After graduation, he went through the traditional interviewing process
with his fellow class members, but when he received no job
offers, he learned that the Robotics Institute was seeking a
part-time engineer to help build a test bed for "inflatable"
robots. At the time, NASA was intrigued by the idea of using
large beachball-like robots to traverse tough terrain and travel
great distances, by rolling and bouncing.

Conceptually, the robot lands on Mars and then the wheels
inflate to twice the size of beach balls. They will go anywhere,
and because of their size, they can travel great distances and
move along in "tumbleweed" fashion, with all of the scientific
instruments stowed safely inside. Scientists on Earth can
decrease the air in the ball so that it will stop to do an experiment. A drill goes down into the ground to anchor it. A proto-
type ball was tested at Carnegie Mellon on simulated Martian
terrain designed and constructed by Heys, rolling back and
forth for hundreds of miles.

Heys, Wagner, Wettergreen, Finch, and a few others had
established this base camp a few weeks ago. Wagner and
Wettergreen were Atacama OPS veterans, but this desert's
stark, breathtaking barrenness is overwhelming for newcom-
ers like Heys, for whom the landscape is overpoweringly surreal. After the base camp had been established, the men picked
up Zoë and dozens of yellow wooden crates of equipment
shipped to Iquique, and hauled their cargo to this site they had
selected. In two days, the tents were up, supplies and equipment unpacked, and Zoë was put back together, which was
when Heys began feeling trapped: He had to face the monster
he created, and now, especially in light of Mike Wagner's
problems in relation to the vision system, he is worrying.

Apprehension that Zoë will have a breakdown in the field caused by something Heys designed, jeopardizing the entire OPS, plagues him. "I had repeated nightmares that once we got to the desert and I saw how rough it was here, Zoë's wheels would fall off."

Heys is not exaggerating the ruggedness of the terrain. In 2003, Waggoner, Wettergreen, Wagner, and a few others had gone off for a day in two Toyota trucks to search out sites for the upcoming OPS with Hyperion. They got the first flat tire an hour into their explorations and used their spare. The second flat came a half-hour later on the second truck, and they used the spare for that. By lunchtime they had a third flat, so they were forced to remove a tire from one of the trucks and balance one end of the truck on a stack of rocks—at which point they decided to take a second tire from the truck they were abandoning and head back to the base camp the fastest route possible before they ran out of tires. That extra spare tire was also needed on their return to the starting point. "One more flat," said Waggoner, "and we'd be trying to hitch a ride on a camel."

Heys is also reeling from the pressure of putting the last-minute finishing touches on Zoë before leaving for the desert. The meetings and debates related to Zoë's design had gone on endlessly, wasting valuable design and building time. "Most people knew what they didn't want. But they weren't quite sure what they wanted in a new robot. They were coming up with wild, impractial ideas." This is typical, he says, of people who deal with software. He calls them "code monkeys" or "eggheads," people who usually know nothing about hardware. The "gearhead/egghead" (hardware/software) rivalry in robotics is subtle but persistent, especially from the engineering side. "They put graphs on whiteboard and scored each configuration," says Heys of the code monkeys. "They couldn't make decisions."

Engineers at the Robotics Institute describe the AI (artificial intelligence) people as more prone to take risks because they devote so much more time to theory than reality and don't fully understand the practical consequences of risk taking. Software folks are geared toward perfection—because theory often works perfectly in simulation—while engineers like Heys learn to live with imperfection. Engineers generally assume the worst case and code writers the best case. As Dilbert said, one engineer told me, "It's the goal of every engineer to retire without being blamed for a major disaster."

Heys had narrowly avoided that major disaster a few weeks before Zoë had been shipped to Chile. The potentiometer, or POT, which communicates the angle of Zoë's axles to Zoë's computers and is a key element in how Zoë will steer—that same mechanism that caused the subterranean rookies such last-minute trouble at Mathies—cracked during testing. The day before Zoë went into the crate to ship to Chile, Heys was working nonstop, machining improved parts to replace it. He worked up to the very minute the crates were shipped. The cracked POT precipitated Heys's nightmares.

Their suppliers had also procrastinated. The aluminum honeycomb chasis, for which they had scheduled three weeks to machine, took six weeks to deliver. And the special lithium batteries electrical engineer Jim Teza had so carefully designed to extend Zoë's life without solar power from two to four hours were delivered two months late.

Heys had used Hyperion as a starting point for Zoë, improving upon Hyperion's design in subtle but significant ways. Hyperion's axles were connected: if the front axle hit a rock or bump, the rear axle moved; if the rear axle collided with a rock or rut, the entire robot trembled. "Now we have body averaging—meaning the chassis is between both axles, and is suspended independently." Both axles steer independently, also. The spine of the robot is designed as a hoop, providing the stability to carry the science autonomy tools. The drivetrain is a compact unit. All bearings and wiring are internal, unlike Hyperion. The hoop, spine, axles are aluminum while the instrument enclosures are aluminum honeycomb sandwiched between layers of fiberglass. Zoë's lightweight (200 kilos) frame will allow it to go four times faster than Hyperion. The aluminum and fiberglass design, says Heys, adds to Zoë's "cool factor."

Heys's position as the lone mechanical engineer is double-edged. He likes making his own decisions and he is aware of the power he has over the code monkeys. But he has never been faced with such awesome responsibilities as a $4 million robot and he worries constantly, knowing that his friends and colleagues are relying on his work.

This is part of the challenge and the benefit of working with and relying on young, inexperienced people. They don't have a lot of trial-and-error history to fall back on; they haven't been permitted the luxury of making mistakes and learning from them, and they have few mentors to consult. And so they often lose confidence and composure. "I really have no one to discuss the hardware with. I feel so much pressure most of the time because I know I am all by myself and I have to make sure everything I do is right—or else."

So far, Zoë was staying together. Nothing bad had happened to the chassis, the wheels, frame, or any of the mechanical parts. "That doesn't stop me from worrying about the moment when everything is going to fall apart. I'm nervous. And," he pauses to momentarily consider his feelings, "at the same time maybe a little bored. I really haven't done much since we arrived here. I have been doing a lot of standing around and worrying."

In contrast to Stuart Heys, who is bored and nervous, Alan Waggoner is very excited about being back in the Atacama for this second year. When we return to the base camp after breakfast, Mike Wagner boots up Zoë and begins to manipulate the robot around the compound in order to investigate its vision problem. But now with the sun high in the intense and striking cobalt sky, Alan Waggoner cannot seem to contain himself. Waggoner is a grizzled, wiry, slightly stooped scientist. His work for LITA will be key, for his organization has been charged with developing the life-seeking, autonomous science instrument that Zoë will carry into the desert. Waggoner's Molecular Biosensor and Imaging Center is world-renowned for developing fluorescent probes and dyes that have played a significant role in the human genome project and are used to analyze gene activity in cells and tissues. This is Waggoner's first attempt to bring fluorescent technology into astrobiology through robots, however, and although he is "fumbling in the dark" at the moment, he is effusive about his work. Ask him about fluorescence, and he gushes with information:

Fluorescence, he explains, occurs when a substance absorbs light of one wavelength and then emits light of a longer wavelength through a probe or a dye. This causes the molecules of a substance to become excited, which, in turn, will either radiate electronic energy or produce vibrational energy, which is what we know as heat and light. The molecule stays in an excited state for a few nanoseconds. When this happens, heat or light—or both—are emitted through the dye and it fluoresces—an activity that can be measured. "This is how we are someday going to find life on Mars," he says. "But first," he added, almost as an afterthought, "we'll begin in the Atacama."

His idea is simple. His team will develop dyes to excite chlorophyl, lipids, DNA, carbohydrates, and perhaps other molecules of life, spray the dye on the terrain, shine intense light on the samples, and measure the activity. Waggoner thinks he has found a way to pull this off with Zoë thinking out the problem and doing the work; the upcoming OPS will be the test for his device. But at this moment, he is not focusing on fluorescence, for the sparse red desert is turning him on.

Waggoner is climbing dunes, one after another, and breathing in the all-consuming vistas surrounding us. "What I like is how you can be all alone in the desert, just go off and stare into space, far away from everyone. This is reward for the politics and heartache in academia, this freedom, however temporary. All the quirks in a person's personality emerge here during fieldwork—even mine!" At one point, he spontaneously drops to the ground and allows Zoë to drive over him, its rubber wheels passing on either side of his prone body. Everyone cheers and Waggoner stands back up, raising his fist triumphantly.

But Waggoner is unaware that Mike Wagner has been woefully unsuccessful in trying to debug Zoë's vision system, meaning that Zoë was not at all under control at the moment Waggoner dropped so daringly to the ground. Zoë weighs 440 pounds. Waggoner would have probably survived the collision with a bump and a bruise, but Wagner was more worried at the moment about the upcoming OPS—or if it would even take place. Right now, Zoë's vision system is a disaster.

The stereo panoramic imager (SPI) is actually a triplet of high-resolution cameras mounted on a pan-tilt unit (PTU) that sits atop Zoë's mast. The SPI is the most visible of Zoë's cameras, the main window for the remote science team to view the desert environment and collect high-resolution stereo images, as well as 360-degree panoramas of the geologic and biologic structures around the rover.

But Zoë navigates through a stereovision system—two additional video cameras, or "nav-cams," located in a less visible location about midheight of the robot, pointing downward toward where the wheels touch the ground. When an obstacle blocks its path, Zoë knows to avoid that obstacle and picks a path to go around it. "But," says Wagner, shaking his head and rolling his eyes, "the nav-cams have to work right."

Wagner realizes that sunlight is more intense in the desert. And since the cameras supplying the data were removed from Zoë in Pittsburgh and packaged individually for shipping, settings might have been accidentally altered. So after discovering the problem, he and Heys repeatedly recalibrate. But the data transmitted from the cameras continue to make Zoë seem almost drunk.

Eventually, Wagner begins to study the images the cameras are recording. "If you look at an object with your right eye, then close that eye and look at it with the left eye, and you go back and forth, opening and closing your eyes, the object you are studying moves and changes in a certain way." But when he tries this through Zoë's cameras, "something looks fishy," he says. Wagner compares the serial numbers of the cameras with the reassembly plan they had followed in uncrating Zoë. Then the solution dawns on him.

Wagner pauses for dramatic effect and shakes his head in wonderment, as if he is asking himself how he could have been so stupid. He is prone to outbursts of shrieking laughter, with which he now suddenly explodes. "The cameras [have] been flipped!" he announces. "The left camera [is] on the right side and the right [is] on the left." Typical robotics riddle and solution. Consult the obvious before going to the extreme. For a moment Wagner is relieved that Zoë's vision problem can be solved so quickly, but the snafu of the switched cameras is a harbinger of things to come.

Wagner dismantles the navigation cameras, reinstalls them in their proper positions, and recalibrates carefully. Then he reboots Zoë, waiting for it to construct a map of the terrain, and then sends it a basic navigation command. And for a moment, the system seems to be functioning smoothly—until Zoë runs into the chair Wagner has placed in its path as a test obstacle.

Wagner then realizes that flipping the cameras is only the beginning of the solution to solving the problem of Zoë's vision system. Something else, God knows what, is wrong. The representations of the terrain, so vital for autonomous navigation, the very reason for Zoë's existence, are alarmingly inaccurate. Zoë is nearly as blind at the end of the day as it had been in the early morning when he had first discovered the problem.

PART TWO

autonomy

CHAPTER SIX

RoboCup

THE LITA PROJECT CALLS FOR AUTONOMOUS SCIENCE and autonomous traverses. But as I had already discovered, the term and the concept of autonomy is a tenuous and difficult-to-define condition. Although robots like Nomad, Spirit and Opportunity, Hyperion, and now Zoë have been hyped as being autonomous, there are, if the word is to be taken by its literal definition (able to make decisions and act on them as a free agent), no robots operating today in hospitals, factories or museums, deserts or distant planets—anywhere—that are truly autonomous.

Many robots are teleoperated, in which a human sees the world through the robot's vision system or camera eye and, through use of a joystick, directs its movements. When Whittaker's robots first crawled into the dangerous depths of Three Mile Island, they were teleoperated, a method which remains an anchoring element of human/robot collaboration. While unraveling the puzzle of its vision system, Mike Wagner is teleoperating Zoë with its joystick. Telepresence, a step beyond teleoperation, enables a person in one location to perform a task through a robot by manipulating instruments and receiving sensory feedback.

Robonaut, a humanoid designed at NASA's Johnson Space Center in Texas to take the place of human astronauts, offers a sophisticated example of telepresence: "A master-slave relationship," says NASA, "whereby the operator's motions are essentially mimicked by the robot." In other words, a virtual-reality display enables the human to see, hear, and feel what the robot is experiencing and duplicate the motions required by the robot to complete a given task. The human and the robot are in a state of oneness.

But oneness is not autonomy. Not even semiautonomy. There are a few robots that can perform narrow, specified tasks or missions autonomously, but even those are not autonomous in the way human beings are. This is the consummate challenge in the robotics world: Creating a robot that will think and act on its own—as close to being "human" as possible. Real autonomy—the actions we humans take for granted—is an amazingly remarkable evolutionary accomplishment.

Think about what are you doing at this moment. Sitting in your office or living room, reading the words in this book, comprehending what they mean. As you read, you might jot down a few notes, lean back in your chair, and cross your legs in contemplation. Suddenly, you hear a noise and glance up. It is your partner with coffee from Starbucks; you recognize the familiar green and black logo. You lift the cup to your lips and sip, feeling and tasting the hot, bitter brew. In a while, you might turn away from this book and reach for another. Or take a telephone call from a friend or walk down the street to make a bank deposit. All of this is normal, just what human beings do that make us who we are. Nothing special.

To the contrary. A human being is the most sophisticated system in the universe—more complex than the largest galaxy. If you linked together all the computers in the world, you would not have a machine that would duplicate the actions of the human brain. Your body is similarly unique. Take the action of grasping the coffee cup, lifting it to your lips, and drinking. There are twenty-seven bones and dozens of muscles and tendons in each hand, perfectly synchronized with the muscles in your forearm—separate systems working in harmony. Pulling yourself out of the chair and walking across the room entails the cooperation of dozens more bones and muscles and the interaction of countless nerves and connective devices from ankle to knee—biological sensors and actuators that control them.

Roboticists have an awesome task ahead of them; they are making progress in their quest for autonomy—often in unusual, unpredictable ways. In his keynote address at the October 2000 Earthware Symposium, the late Nobel Prize laureate Herbert A. Simon—a founder of artificial intelligence and known to stay indoors and tap incessantly at his keyboard—was eloquent and exhilarated about one unique athletic competition.

Here around CMU we have been amazed, amused and gratified and instructed by the developments in robot soccer. For four years and with rapidly increasing skill, computers have been playing a human game requiring skillful coordination of all the senses and remote capabilities of each player, as well as communication and coordination between players on each team, and strategic responses to the moves of the opposing team. We have seen in the soccer games an entire social drama played out with far less skill (thus far) than professional human soccer, but all of the important components of the latter clearly visible.

Here we see in a single example a complex web of all of the elements of intelligence and learning that AI has been exploring for half a century and a harbinger of its promise for continuing rapid development. Almost all of our hopes and concerns for the future can be examined in miniature in this setting.

Manuela Veloso, then a forty-three-year-old mother of three and associate professor, was in the audience that day and surprisingly was the primary recipient of Simon's praise. I say "surprising" because Veloso is a powerful and prestigious voice in a male-dominated field and because, like Simon, she is far from athletically inclined. In fact, she had never played soccer or paid any attention to it—until Hiroke Kitano, a Sony Corporation senior scientist and Robotics Institute alum, came up with the idea of programming robots to compete in soccer. Suddenly it all came clear. Soccer may be the best and fastest way to achieve autonomy.

Humans are autonomous in our everyday lives, prepared to do what is necessary to survive in an unstructured environment, said Veloso. Autonomy for robots means locomotion—the ability to move—in consort with the senses, like touch, hearing, and sight. Vision is a key component of being autonomous. But while many robots can see with remarkable clarity through sophisticated digital cameras, their ability to perceive and react to the intricacies of their environment—to navigate—is limited. Robots can tell the difference between certain colors, but shadows limit their perceptive abilities and multicolors may confuse them. Robots exhibit a minimal capacity of recognition. A robot may perceive that an object has been placed on a table, but will not know, for example, whether that object is a pizza or a pistol. It won't know what a table is, either.

Autonomy also includes an energy source. A human being can go without food or water for many days and continue to function. Most robots need to be recharged or refueled every few hours of operation—or less. But many roboticists have been increasingly excited by the potential of RoboCup and have become deeply involved: These gizmos can think, to a limited degree, like human beings. "They have cognition," said Veloso. "They can do complicated computations without human intervention."

The long-range objective of RoboCup, according to Manuela Veloso, is brazen, ingenious, and nearly within our grasp: To create a team of robots that will play head-to-head, on the same field, with the same ball, following the same rules and regulations, against humans (the World Cup soccer champions, in fact, on or before the year 2050).

Manuela Veloso is unobtrusive—simple, conservative appearance, straight black hair streaked with gray, rimless glasses with black stems balanced on the crook of her nose below a pensive brow—until she begins to speak, at which point she becomes, to say the least, commanding. Born in Portugal and educated at Carnegie Mellon—an electrical engineer with a PhD in artificial intelligence—Veloso speaks with a distinctive accent and a twisted syntax, but her passion is contagious. Australian roboticist Brett Browning, now at Carnegie Mellon working under Veloso, described her as "a human cyclone. She enters a room, interacts with people, and they start feeling motivated. Then she leaves and people feel suddenly turned on and slightly stunned and don't know why."

Veloso shares Red Whittaker's charisma, and although she clearly understands the importance of "hype" and "spin" in gaining media attention for her robots, she is not, like Whittaker, personally motivated by the limelight nor the challenge of the impossible. She eschews the big-leap-forward idea that drives Whittaker and, like Wettergreen, is engaged in the process, preaching incremental steps and universal cooperation within the robotics field.

Ashley Stroupe, who earned her PhD in robotics at Carnegie Mellon, first spotted Manuela Veloso on a *Scientific Frontiers* TV show hosted by Alan Alda early in RoboCup development, when Veloso's little robots were playing soccer on Ping-Pong tables. "Manuela was screaming at the robots," said Stroupe, "and Alda, surprised and amused, told her, 'They can't hear you, you know?' " Veloso didn't care one way or another; she screamed even louder.

Veloso's scientific objectives are perhaps tempered by her perceived responsibility as a member of a significant minority group in the world of technology. A report by the National Council for Research on Women noted the minimal increase in women in science over the previous twenty-five years. One in five women received undergraduate degrees in physics and engineering—only a 10 percent increase. Women in computer science had actually lost a lot of ground. In 1984, 37 percent of all undergraduate degrees in computer science were awarded to women. In 1999, only 20 percent of undergrads earning degrees in computer science were women. In a 2005 article in *Science* magazine, a group of women scientists, including chancellors and provosts of universities and colleges, reported progress, but said continued bias and a campus climate that undervalues women's contributions made women feel "not respected." According to *Science*, women encounter four broad challenges in seeking tenure and opportunities to advance to senior faculty and administrative positions:

> THE PIPELINE: In engineering and the physical sciences, fewer women are trained to the PhD level and encouraged to pursue academic careers. In the biological sciences, the percentage of women professors is inconsistent with the percentage of women obtaining doctorates.
>
> CLIMATE: "Many women attribute their exit from the academy to hostility from colleagues and a chilly campus climate."
>
> UNCONSCIOUS BIAS: Even individuals who believe they are not biased may "inadvertently" discriminate.
>
> BALANCING FAMILY AND WORK: "The responsibilities for family caretaking (for children and aging parents) continue to fall disproportionately on women."

Carnegie Mellon has been one of the few universities to successfully recruit an increasing number of women undergraduates in computer science, however. The faculty has revamped curriculum to be more women-friendly, sponsored a series of summer institutes for high school computer science teachers to discuss gender equity issues, and set up a support system and mentoring program for undergraduate women recruited into the program. It seems to be working. Women made up 8 percent of Carnegie Mellon's 1995 computer science class of 110 while 37 percent of the 130 in the class of 2001 were women. However, since the dot-com crash in 2001, the percentage of women in computer science at Carnegie Mellon has dropped slightly. In 2005, there were 31 women enrolled overall in the class of 141—about 30 percent.

Kristen Stubbs, a PhD candidate at the Robotics Institute, is, as she puts it, "a lone female in a sea of men." Male students sometimes "say stuff in front of me that they wouldn't say in front of their girlfriends, but there has never been a time when I have been excluded because I am a woman." The male atmosphere can be inhibiting and annoying, however, and, Stubbs says, "the fact that I am a woman does make me a little bit more hesitant to say things."

Ashley Stroupe points out that six members of her class of eighteen PhD candidates in the School of Computer Science were women, all in robotics, because robotics is more interdisciplinary and multifaceted than any other computer science area. She believes women have a harder time focusing on a single thing—like programming—and are invariably more comfortable working with people and being part of a team, an unlikely by-product of writing code, she explained. Upon graduation, Stroupe joined the Jet Propulsion Laboratory as an engineer analyzing data sent back from Mars by Spirit and Opportunity and was eventually promoted to rover "driver"— a process of teleoperation through keyboard commands, which is "like trying to drive a car by writing a computer program." Stroupe is the first woman to drive on Mars.

Veloso, promoted to full professor in 2004, has been an inspiring role model to Stroupe and many others, but her position and prestige are reminders of how far behind women are, rather than how much they have achieved. In a faculty of 148 in the School of Computer Science, twenty-nine are women. Of the thirty-nine full professors, Veloso is one of six women.

In work situations, "I never think of myself as a woman. But I notice that I am sometimes more emotional than my man students and colleagues." It is, perhaps, the fact that she is willing to be so emotional and genuine that enhances her popularity and the appeal of RoboCup.

Manuela Veloso and Hiroke Kitano, along with graduate student Peter Stone, conceived of RoboCup not for the sake of the sport of soccer or to win a game or capture a championship flag or to do something no one else has ever done, but to speed the evolution of robot science. They blamed colleagues who worked independently and in secret, hiding their data, for the slow progression toward achieving autonomy. Together, they hatched the RoboCup plan: Persuade colleagues to build and program robots to function as teams and play soccer against one another. Turn it into a big international event that would attract the important people in the field. Urge teams to share all data at the end of the event. No secrets.

This latter point has been a key to the success of RoboCup. "The goal of making a single autonomous creature to think, act, and perceive will never be accomplished by a university or individual working alone. Collaboration of many people will be required. Some will work on a robot's emotion, some on vision, some on artificial muscles. This is the ultimate teamwork project," Veloso said.

The claim that robots could defeat the best human beings in the world at a game devised by human beings and perfected for centuries was, perhaps, outlandish, but when respected scientists go public with such a thesis, it makes news, attracts attention, creates a buzz, and, in this case, provides an opportunity to communicate realistic information to the public about robots. This is the philosophy behind RoboCup.

Many people have a skewed picture of the capabilities of robots. Blame the entertainment industry for the hype and exaggeration, and for America's misinterpretation of what robots are all about. While it is true that Japan's Astro Boy, C-3PO from *Star Wars*, and Robby the Robot from *The Forbidden Planet* are lovable and heroic, *The Stepford Wives*, or Steven Spielberg's frightening and depressing *AI*, depict robots as clones virtually indistinguishable from real human beings, except for an inability to experience the human emotional condition. The 2004 incarnation of Isaac Asimov's *iRobot* with Will Smith portrays a world of robots three decades from now in which robots are corrupted by men, who, in turn, have been corrupted by robots.

This "robots taking over the world" scenario did not begin with Asimov, however; rather, it has been replayed in different forms over nearly a century in literature. The word "robot" was coined by the Czech playwright Karel Čapek in his play *R.U.R.*, produced in 1923. The play is set in a factory of a company called Rossum's Universal Robots that manufactures artificial slaves designed to relieve humans of the drudgery of work. At first, the slaves are primitive, but they soon become more sophisticated and numerous. Gradually they begin to outnumber their masters. After a while, they are taught to be soldiers, fighting wars on behalf of humans, and then, inevitably, they revolt, wiping out the human race.

Americans know that Asimov and Čapek were writing science fiction and that robots will not soon be on the verge of taking over the world, but at the same time we also remain reluctant to fully accept robots as a "friendly force" as do the Japanese; Americans are not quite ready to integrate them into society. But, of course, technologically speaking, robots are not quite ready for prime-time integration, either.

This is the confounding dilemma of robotics: Robots are so unique and impressive, hyped as a novelty by the media at every opportunity, consistently employed by filmmakers as a vehicle for projecting into the future, that people have been led to believe that robots are more advanced and capable than they are in reality. The transition of robotics into the mainstream of American life is happening, but ever so gradually. And despite Manuela Veloso's impact and appeal, it is not beginning on the soccer field. The slow proliferation of robots is, rather, beginning in the home.

Roomba, a robot created by iRobot, a company started by researchers at MIT's Artificial Intelligence Laboratory, including Rodney Brooks, a Robotics Institute alumnus and director of the MIT Artificial Intelligence Lab, crawls around the floors of millions of American households, sucking up dirt and debris. Although it is often called a "housecleaning robot," Roomba's activities are basically limited to vacuuming. Shaped like a flying saucer, Roomba is an efficient little tyke, with the ability to recharge its own batteries and sense through sound waves dirty areas that require extra vacuuming effort. Other companies, Electrolux and Samsung among them, have introduced their own versions of the miniature robotic vacuum. Late in 2005, iRobot introduced the Scooba, a companion to Roomba, which cleans and scrubs tile and wooden floors, and Dirt Dog, a clean-up robot for workshops. Roomba, Scooba, and Dirt Dog are models of the way in which robots are beginning to make a dent with wary Americans—a one-task robot, selling at a reasonable price ($200–$400) rather than the "universal robot" idea floated by roboticists in the 1970s and 1980s—the philosphy and approach adopted by the Japanese.

Robots are also making rapid inroads in the toy and entertainment world, an initiative championed by Carnegie Mellon's Illah Nourbakhsh, who compares the current state of robotics to the early days of the personal computers, beginning with the impact of Apple Computer founder Steve Jobs. Making robots accessible to young people will open the door to the creation of ultrasophisticated humanoids and rovers just as the Apple II made computers more accessible and indirectly launched the digital revolution. Nourbakhsh has introduced and taught a class for high school students at the Carnegie Mellon West campus on the NASA facility in Ames.

Based on a NASA rover, Nourbakhsh led students through the stages of building a TRIKEBOT, a robot he describes as "IKEA's answer to robots"—pieces of plastic and wood, held together by screws, simple to build and rugged for indoors and outdoors. They experimented with ways in which students and robots could live side by side. The students took the robots home at the end of the course.

Nourbakhsh was in his mid-thirties, wiry, balding, and intense when I first met him in 2003. He was the creator of Sage, a mobile multimedia exhibit robot that wandered Dinosaur Hall in Pittsburgh's Carnegie Museum of Natural History's exhibit area, providing video and audio enhancements for museum visitors. Sweet Lips, a similar museum tour-guide robot, plugged itself in, e-mailed Nourbakhsh when it felt sick, and worked peacefully around people; Sweet Lips and Sage performed their tour-guide jobs, avoiding collisions with visitors for nearly five years before being retired. Raven, a similar Nourbakhsh tour guide, is still going strong at the National Aviary.

Neither Sweet Lips nor Sage nor Raven have feet. They move around on four wheels. Nourbakhsh would like to develop a new robot with a foot and a leg and create a hopping humanoid. "Why must a robot walk with two legs? Must a humanoid be *exactly* like a human?" he asked. Airplanes are inspired by birds, but Boeing 707s don't have feathers, lay eggs, or flap their wings, Nourbakhsh says. Humanoids should be similar to humans, he added, in order to help people become more comfortable with them.

As he described his vision of the one-legged humanoid revolution back then, he leaned forward with earnestness and excitement, positioning his face directly into my face and talking rapidly until he was almost literally pressing into me. Then suddenly, he sprung up, as if he were on a pogo stick. In fact, within a few seconds, he *is* on a pogo stick, although a considerably oversized version, what he called the BowGo, which enabled him to leap to the ceiling and literally fly across the room, down the hall to his secretary's office and back, like Spider-Man.

The BowGo is a toy he was attempting to market. Nourbakhsh's "toy initiative" was behind the PalmPilot Robot kit, which turns a PalmPilot into a four-wheeled robot and has triggered a desktop robotics movement—little mechanical creatures that manipulate items (for fun or utilitarian purposes) on an executive's desk.

In December 2004, the *New York Times Magazine* ran an in-depth feature story about the year's high-tech hit of the American International Toy Fair, a barrel-chested, fourteen-inch-tall, walking, talking $99 humanoid called Robosapien, which will dance, sing, "guard your closet, keep your cat off your desk and pick up your socks." Robosapien is definitely a mechanical tour de force, with many transistors, motors, and dozens of sensors, but it is mostly teleoperated. Out of the box, it can't think or act on its own. It lacks what Wettergreen is seeking for Zoë and Veloso wants to give to her soccer team: autonomy.

Roboticists often measure autonomy in loops. Roughly speaking, opening and closing a "loop" means that a robot will initiate and complete an action, independently. A broad loop—more complicated—is indicative of a higher degree of autonomy. Spirit and Opportunity are examples of a narrow loop, indicating a lesser degree of autonomy. "At JPL," said Wettergreen, which is where the MER team is located, "people look through the 'nav-cams,' see the scene, decide where to go, then communicate that path through a series of precise commands back to the robot. The robot closes its eyes and follows orders until each command is completed. Afterwards, it opens its eyes and looks around and says, 'Did I get there?' " Wettergreen was describing essentially what Ashley Stroupe is doing—driving the robots. This is a very limited level of autonomy.

Wettergreen added: NASA announced that a few months after landing, Spirit and Opportunity had navigated ten meters autonomously. What they meant by "autonomously" was that they allowed the robots, periodically, to open their eyes and see the terrain directly in front of them through another set of cameras closer to the ground, so that, as NASA had announced, when the robots encountered a rock, they could go around it and continue on the designated path. This was a higher degree of autonomy and the second way the MER robots close their loop. The meaning of autonomy has evolved as technology has evolved but the term remains very ambiguous.

ILLAH NOURBAKHSH'S TOY initiative and his comparison of the robot with the Apple II was quite apt: Computers today symbolize the basic anchor and foundation of the world as we know it, in every conceivable discipline, introduced by the postwar generation but championed and now embedded in children and young adults. Robots are an inevitable extension of the computer age, as the growth of the Robotics Institute symbolizes. Robotics departments at major universities are cropping up throughout the United States and around the world, especially as degree programs like Carnegie Mellon, MIT, and Stanford continue to educate and produce young roboticists. High schools and middle schools are initiating robotics programs because, as Nourbakhsh has correctly postulated, robots can capture the imagination of young people and lure them into science and engineering much more so than a textbook or a teacher.

Nourbakhsh's success with his toy initiative and the courses he designed eventually led to an offer of a prestigious full-time position at NASA Ames in 2003 to focus on the development of robots that could be partially controlled by people and partially autonomous—a concept called "collaborative control." In the future, human crews on missions to Mars or the Moon will be limited to small teams. Astronauts will need robot helpers to do much of the work—with humans supervising and acting as quality-control officers. Nourbakhsh returned to Carnegie Mellon in 2005, introducing a new program called TeRK (Telepresence Robot Kit) to involve young people in the robotics movement. TeRK includes the development of affordable robot designs that can be built with hand tools and commercially available off-the-shelf parts.

EQUALLY IMPORTANT TO the toy initiative idea, and perhaps more ingenious, are the seemingly impossible pie-in-the-sky applications and schemes that attract attention from every direction simply because they are so unique—and at the same time prove to be serious scientific and/or technological endeavors. RoboCup is one and it almost immediately precipitated a major change of approach in the way in which research in robotics was conceived and funded.

The first RoboCup international, supported by Sony, debuted in 1997 in Japan, with forty teams representing a dozen countries. In 2005, again hosted in Japan, 330 teams representing thirty-one countries competed in a nine-day competition in front of hundreds of thousands of fans—the largest RoboCup ever. Germany hosted RoboCup 2006, in concert with the World Cup, also in Germany, and perhaps a harbinger of things to come.

This mass of public interest and support was also DARPA's vision of future robotic evolution when in July 2002, DARPA (Defense Advanced Research Projects Agency) announced the Grand Challenge—a $1 million winner-take-all competition in which a robot was to drive on its own, racing against other robots, from Los Angeles to Las Vegas—250 miles—within twelve hours.

DARPA manages and directs selected basic and applied research and development projects for the Department of Defense, and pursues research and technology where risk and payoff are both very high and where success may provide dramatic advances for traditional military roles and missions. Carnegie Mellon's interest in pioneering concepts and "terrestrial test beds" has made DARPA a significant source of financial support over the past decade.

Such autonomy in a robot—as well as DARPA's idea of a robot dramatically dashing across the desert—was exciting and intriguing. Charles Lindbergh going across the Atlantic in a single-engine plane or Columbus sailing from Spain to the New World may have had better chances for success, many roboticists agreed. But few details were available about the DARPA competition when announced, so there was very little buzz around the Robotics Institute about the DARPA Challenge—until some months later.

Other scientific competitions may have been inspired by the RoboCup example, however, including the recent Virgin Galactic reuseable spaceship funded by Sir Richard Branson, which won the $10 million Ansari X Prize. Its designer, Burt Rutan, backed by Microsoft billionaire Paul Allen, spent more than $25 million to beat two dozen competitors in this developmental sweepstakes. A venture capitalist has ordered five of these spaceships. In the first eighteen months he intends to make a fortune by sending into space 700 customers who will pay $200,000 each for the ride of their lives.

CHAPTER SEVEN

The Color of Thinking

WHILE MANY TEAMS DESIGN AND BUILD ROBOTS FOR RoboCup competition, Manuela Veloso considers the Aibo dog ($1,500–$2,000), right out of the box from Sony, the perfect RoboCup robot. Like any dog, beagle-sized Aibo will sing and dance, wag its tail, voice pleasure when you pet it, or act out its displeasure by shaking its head and refusing to listen to commands. But it can also chase a ball, trap it with its paws, and kick it with its rearing, blinking snout. There are many other RoboCup competitive classes of robots, somewhat smaller than Aibos and considerably larger, as well. Most recently, Segways, unicycle-like independent people movers at a little less than 100 pounds, have been programmed and redesigned to play soccer and compete in the RoboCup games. This too is a Carnegie Mellon initiative. But the Aibo RoboCup competitions have been the most popular.

The American Open at Carnegie Mellon in May 2003, which included RoboCup teams from the United States, Canada, Mexico, and Central and South America, constituted Veloso's next level in spreading her RoboCup agenda: To introduce regional events, making the competition more accessible to a larger, more diverse and less specialized constituency, including university and high school students and teachers to whom the cute Aibos and other small robots were precious, accessible, precocious, and alluring. (The American Open has since morphed into the United States Open while other countries in the Western Hemisphere are coordinating their own events— warm-ups for the annual international RoboCup.) At the American Open, the students also came to see a very special guest, the most distinctive and impressive character in the robotics world at that moment: Honda's Asimo (Advanced Step in Innovative Mobility), the world's most sophisticated humanoid. But the Aibos took center stage at first.

Color coding is the key to how Aibos and other RoboCup robots think and see. In their world, the ball is orange. The goals are yellow and blue. There are markers of varying pastels. "That's their map of the world," Veloso said. When the Aibos line up on the field, three in front, supported by a goalie behind, they rear their heads like stallions. They are looking around; their small video cameras only have a narrow scope of vision, so their heads are always moving, to find the colored markers and localize. To put it another way, the robots are in the middle of an ocean, and each color is like a lighthouse, helping them determine their current location and where they next need to be. They don't map in the same way as Groundhog; the color signals and informs them. Aibos have many sensors—for distance, acceleration, and vibration. They have paw sensors—not to mention a speaker in their nose and a stereo microphone.

But Aibos are very delicate mechanisms, easily distracted, almost like puppies. At an American Open practice session, some of the visiting students began flipping the light switches, totally disrupting the robots' vision systems. Veloso's students worked through the night to recalibrate their little dogs. Perfect calibration is critical for Veloso's robots to achieve her objective:

"I have a ball in front of me—what should I do—pass the ball? Kick? Dribble? What's the probability of scoring? What's the probability that if I shoot directly at the goal, the ball goes through?" Veloso explained. These ideas will filter through the robots' brain—not as humanlike thoughts, but as bits of information that initiate certain deductive algorithms or formulas. "My robots think in a loop. It is always going through their minds—where is the ball, where are my teammates, where am I? The algorithms we develop permit them to answer these questions."

Like any soccer player, the robot knows that the objective is to win through their own efforts, combined with their best assessment of how to work with others. They even seem to know when they are winning and what they need to do to stay ahead. "If they are winning," said Veloso, "and it is near the end of the game, then the robots move into defensive modes."

"A coach doesn't tell them to do this?" I asked.

"There are what we call 'locker room agreements,' built into the programs. People tend to think that programming a robot makes them do things exactly the way they are programmed to do it, with no variation, but our program can be flexible. The range of situations robots can respond to is enormous."

Theoretically, the RoboCup robot is in perfect harmony with its teammates. Each robot is first an individual system, but also shares a combined vision, multiplied by the number of robots on the team. Thus, Veloso's loop is an exercise in cooperative sensing—a robot mind-meld, so-to-speak—or "multiagent" in roboticist language. If you are a soccer aficionado, or if you are turned on only by Schwarzenegger-style fast-paced action, or if you are impatient with dawdling bunglers and seemingly clueless buffoons, then RoboCup is not for you. Wait until the year 2050 when sophisticated bipeds go head-to-head with humans and beat their pants off. Today, precision and picture-perfect play is not yet technologically possible. But if you're a patient spectator, aware that these robots are thinking for themselves, in Veloso's loop, get ready to be amazed and impressed. Throughout the match, the Aibos trip, fall, get tangled up with one another, and go plain crazy at inappropriate moments—sometimes in the middle of the heat of the action—chasing their own plastic tails.

"You need to understand," Jim Bruce, one of Veloso's grad student disciples told me, "what an awesome challenge it has been just to get them to play the game at all." In other words, you have to recognize the magnificence of the feat: Little toy dogs are playing soccer all by themselves. Unlike Spirit and Opportunity, Hyperion or Zoë, no human is sending them orders.

Bruce has participated in four international RoboCup competitions so far and has four trophies to his credit. Each year he has seen how much more sophisticated the robots and the games have become. One year, Australian roboticists programmed the Aibos to walk on their elbows so that their forelegs could block and shoot. This also brought the dogs closer to the ground so they could propel the ball with a combination belly flop/chest maneuver instead of just pushing it with paws or nose toward the goal. Later, the University of Pennsylvania figured out how to make the Aibos perform a forearm kick. Then the Germans refined a bicycle kick. Carnegie Mellon developed a header. Code to all of these innovations has been shared and subsequently fine-tuned, so now all of the Aibo teams benefit from the innovations of their competitors.

Before the start of the match at the American Open, Veloso introduced her team—the humans, not the robots. These are very nerdy-looking kids, mostly boys with ponytails and sandals, constantly tapping away on laptop computers, grunting and making faces at one another and laughing, even while Veloso announced their names into the microphone. They were getting Super Bowl–style introductions, and the crowd was into it. People were hooting and yelling each time Veloso said a name. Her students, however, kept on typing into their laptops, oblivious to it all. Veloso was very laid back. She spotted her dentist in the crowd and introduced him, as well.

Tucker Balch, the professor leading the Georgia Tech Yellow Jacket team, Carnegie Mellon's first opponent, followed Veloso with his own opening remarks. He introduced and commended his students, and pointed out that both teams were of Carnegie Mellon heritage. Balch, a former F-15 Air Force pilot who wrote his dissertation under Veloso at Carnegie Mellon, revealed that, in the spirit of the RoboCup movement, Veloso voluntarily gave the Yellow Jackets the software her students had developed the previous year, thus providing his team with a gigantic jump start. The code is contained on memory sticks slipped into slots in the Aibo's abdomen. "We have worked very hard to improve this code," Balch told the standing-room-only crowd of about 400, squeezed into a converted ballroom in the Student Center, ringed with bleachers, "so you will see some differences between the two teams—especially in passing—which is where we focused attention—if it works right." Balch's Yellow Jacket team had also tweaked the code so that Aibos would chase the ball and swarm, hockey-style, into the boards in the corner of the green felt playing area, ten feet wide and fifteen deep, to trap the opposing players and retrieve the ball.

The play-by-play announcer, another geeky student, retrieved his microphone and announced that the Georgia Yellow Jackets, the red Aibos, would kick off to the blue Carnegie Mellon Aibos. He counted down, "three, two, one—go!" and the match was under way. The action was, as I said, surprisingly riveting.

First, the ball was pushed toward the blue goal, but then there was a tussle between a red and a blue dog. Since the Aibos have plastic paws—no hands—the ball tends to roll easily and often in varying directions. Now the ball was returned to the middle of the field and there was another fight for possession, but when the ball squiggled behind the scuffling dogs, none saw it for a few long seconds. The crowd was yelling, "Behind! Behind!" It is easy to forget that they can't understand a word you are saying.

A Carnegie Mellon dog retrieved the ball and sent it with a mighty nose shove across the field toward the Yellow Jacket goal. The average Aibo nose shot does not have much juice; the ball tailed off the mark and rolled to the side of the goal and into the corner. The Yellow Jacket goalie knew enough to stay in his position and wait for a blue Carnegie Mellon Aibo to get the ball. Which it did: another nose shot pushed the ball toward the goal, and once again, it tailed off and narrowly avoided a score. This time, the goalie followed the ball, but couldn't seem to move it out of the corner. A blue dog appeared. They fought for control.

Then came a play so painfully typical of Aibo RoboCup competition—simultaneously brilliant and exciting and also depressingly pitiful. The blue Carnegie Mellon dog wrestled the ball away from the Yellow Jacket goalie and, using a neat belly-flop maneuver, sent it across court and right toward the Yellow Jacket goal. The ball stopped about an inch from the goal line.

Two teams of Sony Aibos in a four-legged RoboCup match.
Photograph courtesy of CORAL Research Lab, Veloso, Carnegie Mellon University.

Now the crowd was really into it. Whatever hesitancy these fans, mostly high school and college students, teachers and parents, may have harbored against RoboCup, they were now deeply into the competition: *Ooooohing* and *ahhhhing* and screaming at every opportunity. Manuela Veloso was scream-ing the loudest and yelling instructions.

The excitement was enhanced further when another Carnegie Mellon blue dog suddenly appeared and headed for the ball. Everyone in the bleachers and crammed on the floor, pressed together like sandwiches, knew that this was an inevitable Carnegie Mellon goal. They could feel it. All the blue Aibo had to do was touch the ball with its nose or prod it with its paws—or "breathe on it, for God's sake," a fan yelled in encouragement and confidence. The crowd was going crazy and, sensing the obvious, began to break out in cheers.

A team of Aibos lining up to do battle. *Photograph courtesy of CORAL Research Lab, Veloso, Carnegie Mellon University.*

And indeed, the Carnegie Mellon robot reached the ball, grabbed it with its paws, hesitated, and then, suddenly, completely illogically, whirled around in the opposite direction and shot it the wrong way! The crowd shrieked its surprise and disappointment.

The game went on that way, with the ball moving back and forth across the field without a score: Many tussles and fights for the ball; many shots toward the goal which either miss or weren't propelled with enough steam and eventually tailed off. Periodically, the robots got tied up with one another, triggering an official time-out. Sometimes the goalies got tied up in the corner boards of the goal—forcing another time-out.

When the Aibos committed a foul, they were put into a penalty box for thirty seconds. While missing one of its four players was a hardship to a team, the real challenge from a

Goal! An Aibo RoboCup match. *Photograph courtesy of CORAL Research Lab, Veloso, Carnegie Mellon University.*

computer software point of view was how would the robot adjust when it was dropped back into the action. Robots don't have a memory like humans. The robot doesn't know it was playing soccer and it hasn't been watching anxiously from the penalty box; it was totally unaware of what has been happening in the game. This scenario, when a robot is dropped into the middle of a mysterious world and must determine how to adapt and proceed effectively, on its own, whether on a soccer field or in the middle of a coal mine, actually has a name: The Kidnapped Robot Problem. Meaning that, in essence, the robot is expected to overcome surprise and uncertainty—and reorient itself.

There was no score in the first half of the Carnegie Mellon–Georgia Tech contest, but Carnegie Mellon took command early in the second half with a quick goal. Later in the half, there were two significant events. At one point, the Yellow Jacket goalie sent a long, beautiful pass down the field—the kind of pass Tucker Balch promised might occur. However, the Carnegie Mellon goalie skillfully blocked the shot; otherwise, it would have been a Yellow Jacket score to tie the game.

But later the goalie also demonstrated the terrific Yellow Jacket passing code, by passing the ball to one Aibo, who passed it to another Aibo, who took a mighty belly-flop shot. Unfortunately, it missed, but it was a beautiful play, just as the code writers intended, unmarred by inaccuracy. No other move throughout the game so clearly demonstrated the potential of what could someday be a real exhibition of skill and timing.

The other significant event of that half demonstrated the total opposite of that experience and represented the distance RoboCup must travel to get to the potential of the promised 2050 achievement. At one point the Yellow Jacket goalie did not see the ball roll behind it. Then it unknowingly backed up into the ball and pushed it into the goal—for a second Carnegie Mellon score. Carnegie Mellon beat Georgia Tech 2–0. Georgia Tech defeated itself, said Balch afterward. "We had a great offense and no defense, but this is the very beginning. Wait until next year," he promised.

Later in the American Open, Carnegie Mellon achieved a one-sided victory in the small-size league. Here, five little lunchbox-sized wheeled gizmos with pastel-colored dots on their lids, built with motors and wheels by students, went head-to-head with their arch rivals—Cornell University's Big Red team. While each Aibo had its own vision system, the small robots saw through a camera mounted on a bar above the center of the field and connected to computers sitting on the tables surrounding the playing field. Unlike the Aibos, the Big Red team was autonomous as a unit and each robot was much faster and could shoot the golf ball they use with great velocity—as fast as 40 mph in some cases.

Beating Big Red was especially important, Veloso told me. For the first two years, 1998 and 1999, Carnegie Mellon defeated everyone in the small-size division. But Cornell, under the leadership of Raffaello D'Andrea, designed a graduate program in robotics around RoboCup by engaging electrical and mechanical engineers with programmers, which has been Carnegie Mellon's strength, but in a more evenly balanced equation. Cornell significantly raised the level of competition by developing more sophisticated hardware, adding wheels and motors to previously stripped-down robots, and inventing special ball-control dribblers which shoot with ferocity and accuracy. Cornell defeated Carnegie Mellon the past three years, winning two championships. This year may be different, said Veloso, although win or lose, Cornell's commitment to RoboCup is an endorsement of her vision.

Now, a few seconds before the Big Red game, Manuela Veloso's arms were around her team members, graduate and undergraduate students, as she whispered strategy to them. She peered over the table lined with laptop computers, stretching her body toward the gizmos on the playing field, chewing on her knuckles. Her eyeballs were popping with tension. The robots lined up. The announcer counted down "Three, two, one."

Cornell immediately took control of the ball, but suddenly, a Carnegie Mellon robot streaked forward, stole the ball, drove down the field, and scored. Later, at the end of the first half, with the Carnegie Mellon team leading 9–0—an incredible, unbelievable upset, Jim Bruce said: "We have to make some adjustments during halftime."

"Don't you dare," said Manuela Veloso, laughing and poking him in the stomach. "Don't change anything—for once, this is perfect."

THAT DAY, AS WE watched those soccer-playing bots scamper around, a student engineer remarked out of the blue: "One thing that I'm not sure about in this whole RoboCup deal is the 'human factor.' Do great human soccer teams dominate because they are faster, more disciplined, more skilled than others? If so, then we will eventually build machines that can beat humans—it is inevitable. But if they win through guile or instinct or one of those unquantifiable things that are really at the core of intelligence, then I doubt it will happen . . ."

I presented this thesis to Veloso, who partially dodged the question: "Whether they will win is a different question, but robots will be able to play against humans, I think, in twenty or thirty years—and compete fairly evenly. We will make that much progress. Of that I am certain."

"Why would humans even want to play against robots?" I asked. "Seems like a lose-lose situation."

"Because it will be challenging. Didn't a computer beat Kasparov in chess, and he didn't have anything to gain, either, did he? He responded to the challenge."

She is referring to IBM's "Deep Blue" defeating World Chess Grand Master Gary Kasparov in a showdown match in 1997. Everyone thought that Kasparov was invincible—even Kasparov, whose frustration was evident. He threw up his hands, shrugged his shoulders, and stomped off the stage in disgust after losing. Later, when asked how he felt about being defeated by a machine, Kasparov replied. "Well, at least he [Deep Blue] didn't enjoy the victory."

"The concept—the World Cup champions—best in the world—losing to a bunch of machines—is kind of ridiculous, if you think about it," I said.

"Oh well. Not a bunch of machines—they will be robots. A bunch of machines are refrigerators. Robots are not machines."

"What about human emotions," I said to Veloso. "Do these robots get scared, excited? Do they show emotion and a competitive spirit?"

"Never. They don't get nervous."

"They aren't involved?"

"Those who get involved is [sic] us—watching them play the game. These robots are really our creations; we are competing against our ourselves and our colleagues. They have a lot of creativity in them."

"It almost sounds like you are describing a human being."

"I have been thinking about this problem—robots and humans, you know. And it is probably true that robots will be creatures without a soul. They will lack something inside, something spiritual that we humans have. But we also don't know what a human being is, really. We don't really know if we are creatures of God or DNA. There are people that believe that we have a conscience and a soul and others who believe that we are just matter. And if we are just matter, then probably robots are exactly like we are."

"It is eerie to consider," I said. "Robots will look like us, walk like us. We won't be able to tell the difference between you and me and the robot one day."

"How do you know I am not a robot inside, right now?" she asked, looking at me rather stonily.

"Are you?"

"I am not telling."

Later, Tucker Balch added insight into this human/robot dilemma. "We are many years away, technologically, from the serious moral and ethical problems we will have to confront when and if robots start getting close to us. What I see much closer to today is a robot that enables my mom to live at home for many more years, a robot that can stay with my children, one I actually trust, when I go to watch a movie. Those things are going to happen in the next fifty years and make a positive impact in the same way that the refrigerator and running water improved our lives. Those robots will be subservient to humans."

Balch cited his colleague Ronald Arkin, Regents' professor and director of the Mobile Robot Laboratory. Arkin, says Balch, "recently pointed out to me that we may face ethical issues much earlier. The problem isn't about our responsibility to treat robots properly or to give them rights, but about how the availability of servant robots might change our moral compass. For instance, if robots are portrayed as "slaves" and we treat them as such, this attitude may spill over into other aspects of our society."

Essentially, Balch and Veloso are mainstream roboticists providing a positive point of view of a vital upcoming robotics age. But there's a dark side of the robotics movement, represented by a voice from the past and a living legend whose life's work and obsession reflects a much more frightening portrait of the future. His name is Hans Moravec and, in robotics, he is the world's veteran true believer and a cult figure among science-fiction writers. The self-replicating sentient cyborg in Dan Simmons's science-fiction novels *Ilium* and *Olympos* are called "moravecs."

In person, Hans Moravec seems somewhat robotic, himself. He talks as if he has been activated by voice-recognition software. His body is electrified with awkward, twitching tics and other involuntary movements. He's got a round pink face, a shining bald dome, and an eye that always seems to be halfway closed. In his Carnegie Mellon office, his feet are in and out of his sandals and he pads up and down the halls, barefoot. Dots of spittle accumulate at the corners of his mouth. His wristwatch hangs from a belt loop on his khaki hiking pants.

At Stanford as a graduate student in 1972, Moravec inherited Shakey, a robot developed to go to the Moon and then subsequently abandoned as governmental and political priorities and budgets changed. Shakey had jerky, uncoordinated movements, but it was the first robot to autonomously locate objects and steer around them—and then provide a rationale for doing so. Shakey was eventually replaced by a second-generation rover called the Stanford Cart—also destined to never see the lunar surface. But with Moravec's programming acumen, Shakey could see well enough to navigate from one room to another, following a white line around a colored barrier—although it would take six hours to do it.

Moravec was a living legend even then, and his image continues to loom in robotics lore today, because of his unwavering dedication and commitment. Moravec was so involved in his work at Stanford that he spent six months living in a secret sleeping space in the rafters above his lab accessed by a well-concealed rope ladder. He refused to leave the facility. He bathed in a sink; friends delivered groceries and smuggled in six-packs of beer. His favorite dinner? Cheerios topped with bananas, swimming in chocolate milk—and Budweiser for dessert.

Hans Moravec has been laboring tirelessly on a "universal robot," long convinced that robots will phase out people, sooner or later. "We need an all-purpose robot, a universal robot—a robot who can do everything your heart desires— except perhaps make love and ride motorcycles. This creature would, by 2010, successfully tackle different jobs like automotive repair, furnace maintenance, bathroom cleaning, or home security, simply by plugging in different cartridges of software programs. It will be inexpensive and efficient," he told me.

That's been Moravec's concept from the very beginning. As the decades pass and technology advances, each of four stages of this robot will become more "human" and "moral" until they literally become—us, Moravec continued. The transformation from man to machine will be painless, though quite petrifying.

In his 1988 book *Mind Children*, Moravec describes a robot surgeon peeling a living brain to harvest its data: "Layer after layer, the brain is simulated, then excavated. Eventually, your skull is empty, and the surgeon's hand rests deep in your brainstem. Though you have not lost consciousness—or even your train of thought—your mind has been removed from the brain and transferred to a machine."

Although most roboticists are upbeat or at the very least not as gruesome (or honest), Moravec gets to the root of the resistance to robots, primarily in the United States. Humanoids specifically and robots generally threaten the independent spirit Americans have long cherished, even though the notion of rugged individualism is much larger and more pervasive than the reality.

"Americans are uncomfortable in situations in which we lack control or feel threatened or crowded," Tucker Balch told me. Humans respond more favorably to robots that move at a slower-than-normal human walking pace and don't crowd them. Robots must remain at significant distances. Americans will need to become much more knowledgeable and sophisticated about robots before willingly allowing such mechanical creatures to enter into their daily lives—and space—as partners or helpers.

Human/computer interaction, as an academic discipline, has been viable in the United States for the past three decades, but human/robot interaction is something just now being considered in a serious way by academics. The fact that robots are now able to operate under certain conditions in a semi-autonomous mode has precipitated this field of study because it means that, as technology advances, robots will be interacting more frequently with other robots and, more importantly, with humans. How to accept and negotiate that contact—to overcome the societal queasiness that robots cause—may soon become as challenging as the development of the technology that precipitated it.

CHAPTER EIGHT

Asimo and Friends

W HEN I MET ASIMO (ADVANCED STEP IN INNOVATIVE Mobility), the world's most sophisticated humanoid, developed by Honda Motor Co., Ltd., it was embarked on a nationwide, ten-city "Say Hello to Asimo" tour. An engineering marvel, Asimo (its name is also a play on the Japanese word for "leg" or "ashi") paraded in a figure eight, climbed up and down stairs, played with the crowd by singing, dancing, and joking—and charmed and converted the curious. But for all of its flash and pretension, Asimo, compared to Aibo, exhibited little autonomy; rather it was controlled by technicians manipulating it with Apple PowerBooks behind a black curtain backstage. It was interesting, however, to meet and speak with Asimo, for it is pioneering the wave of the future, according to Jeffrey Smith, then executive vice president of marketing for Honda/U.S., who told me right at the beginning of our conversation to push Asimo.

I looked at Asimo warily. Asimo looked back at me. It was then that I first realized how surreal it is to try to communicate with a robot. At four feet tall and ninety-seven pounds,

Asimo is smaller than I—but it is off-putting, to say the least, and somewhat daunting. Asimo is solid wire, plastic, and aluminum through and through—no skin, no warmth, no DNA, as Manuela Veloso pointed out. This is not like addressing a dog or even a horse, with whom you know you have some natural blood-flesh-bone and spiritual connection. Asimo may resemble a human—and it is also very cute—but the resemblance wasn't close enough to comfort me. Frankly, I couldn't get past Asimo's eyes. You can know a human or an animal by looking deep into his or her eyes, but staring into Asimo was like being on *The Today Show*, staring into the camera without Matt Lauer to guide and comfort you.

I had recently talked with Jody Forlizzi of the Human Computer Interaction Institute of Carnegie Mellon's School of Computer Science. Forlizzi has developed "the Hug," a soft, cuddly robot pillow that uses sensing technology and wireless telephony to provide social and emotional humanlike feedback. The Hug, more concept than reality, is for people who communicate from long distances by phone. So, theoretically, the Hug will allow callers to feel each other's warmth while talking with them, as well as allow them to experience each other's touch through what is known as haptic (from the Greek word "haptikos," meaning to grasp or perceive) interfaces. The Hug will store personal, touchy-feely messages if the calling parties don't make a connection.

Essentially, here's the Hug scenario: You step out to buy a newspaper or go to the bank, and when you return home, your Hug is blinking. So you pour a glass of wine, settle down into your sofa, put the Hug in your arms, and turn it on. Then you get the message, like voice mail, from your mom in Philadelphia or your girlfriend in San Francisco—plus the warmth and the touches that have been stored up for you. Who needs a date, after this—or a spouse, one wonders?

Not that I want Asimo to hug me—and I don't really want to push him. But this is what Jeffrey Smith has told me to do: "You want me to push Asimo? Are you kidding?"

"Go ahead," said Smith. "It will be fine." Smith was demonstrating the fact that Asimo never rests. Sitting or standing, Asimo is always alert, ready for action.

So I pushed. And Asimo pushed back—not hard, but enough to show me that it was a creature not to be messed with. Of course, I already knew this and had no intention of messing with it.

I tried to get on Asimo's good side by introducing some friendly banter, but I was uncomfortable. After all, it was rather eerie to talk with (or to?) a creature who seemed, on the one hand, very human, while knowing all along that it was a machine. "Do you like how I am dressed?" I asked, lamely. I couldn't think of anything else to say. What do you say to a robot? I remember thinking. It's not like you are really talking with someone. "How about those Lakers?" would not break the ice.

Asimo shook its head, no.

"Do you think I should go to the Gap and buy another T-shirt?"

That was another stupid question, but give me a break. I don't have a lot of experience talking with robots. Few people do, except perhaps for a researcher at MIT, Cynthia Breazeal, whose Kismet robot—a talking head—is programmed to recognize emotions in a human voice during conversation and can respond mechanically with sympathetic eye and mouth movements. This is a bit more effective than Asimo, who was now nodding at me, but a long way from emanating a real-life connection.

Asimo, now obviously bored, turned and stomped away, sensors and motors—Asimo has been constructed with 27 degrees of freedom (joints)—whooshing and buzzing, clearly a rejection of my low-level intellectual discourse.

My encounter with Asimo occurred in 2003 during the American Open. But Asimo continues to improve, as Honda pours millions of dollars into its development. Eighteen months after I first met Asimo, it embarked on a second tour of the nation, at which point it could jog (a most incredible engineering feat), pick up small items, tilt its head side to side, and dodge objects in its path with considerably more grace and finesse. The most recent version of Asimo is being prepared for office work. It can push a mail cart or carry a tray of coffee mugs and set it on a table—although it cannot yet distribute the mugs or pour milk and sugar. Equipped with sensors to read microchips in ID cards, Asimo will greet a colleague by name, even from behind. Not that Honda intends to transition into the robot business at the expense of its other income sources. Asimo, which can be rented for a little less than $200,000 annually, is a test bed for developing and improving safety and navigation features for trucks and automobiles.

Sony's Qrio (pronounced "curio"), an "entertainment robot" introduced in 2004, is about half the size of Asimo. Qrio can sing, dance, rollerskate, and pick itself up when it falls. It can, through advanced voice recognition software, even engage in conversation on a bit more sophisticated level than Asimo could when we last met. As an added bonus, it can function as a high-tech *TV Guide* and modified personal digital assistant. Ask Qrio what's on TV, and it will access the Internet and check the TV listings, for example. When the user selects a program, Qrio can point at the TV and, with the infrared emitters embedded in its hands, change the TV channel.

Sony continued to add new and amazing innovations to its humanoid every couple of months. The December 2005 version of Qrio can wiggle its arms and fingers in break-dance fashion. Qrio had also been endowed with something many humans might find helpful—a third ("chameleon") eye on its forehead that enables it to see several people at once and zoom in on one of them. Despite such advances, early in 2006 Sony suddenly announced that it was discontinuing both the Qrio and Aibo product lines.

I contacted Manuela Veloso after Sony made its announcement, for I realized how much she valued Aibo. Veloso is well-connected with Sony through her long affiliation with Hiroke Kitano. She said that she was certain that Aibo would never return—"the end of an era," but that "Qrio might be back. It was never introduced as a product on the market." Meanwhile, Aibo teams will continue to compete in RoboCup until there are no more Aibos available for purchase and the current Aibos are retired. Her students would continue to compete in the small-size league and clear the way for their big brother Segways. But as to the loss of Aibo, Manuela Veloso was very sad.

CHAPTER NINE

Frustration

Y OU CAN SEE MANY STRANGE CREATURES ROAMING THE halls of the Robotics Institute in the middle of the night. I once ran into a young woman who blocked my way and asked me why I was hanging around the campus so late. She was very pretty, but her voice was artificially eerie and her eyes were cold. "It's Lara Croft, of *Tomb Raider* fame," I said aloud. I was talking to myself, just to calm down.

Then Dani Goldberg, a Robotics Institute postdoc, stepped out of the shadows and put his hand on the robot's shoulder. "Don't worry," he said, laughing. "It's Grace." That was my first introduction to Grace (Graduated Robot Attending a Conference).

Grace, as its name indicates, was being prepared to participate in the Mobile Robot Challenge at the National Conference on Artificial Intelligence, sponsored by the American Association of Artificial Intelligence. Over the past half-dozen years, the AAAI has challenged teams of scientists and engineers to program a robot to act like a legitimate attendee—ride an elevator, stand in line, register, schmooze with friends, present a paper,

answer questions. To this point, no one had come close to suc-
ceeding, but in 2003, five institutions, led by Carnegie Mellon
roboticist Reid Simmons, joined forces to beat the Challenge.

Valerie, Carnegie Mellon's roboceptionist, in her
business attire. *Courtesy of Debra Tobin, Robotics
Institute, Carnegie Mellon University.*

Truthfully, I knew that Grace was not Lara Croft. I was joking with Dani—sort of. At that time, Grace's body resembled
an oil canister and it navigated the hallway on wheels. But for
a split second, running into her in that hallway late at night,
alone, I was flummoxed. The way she talked—the direct manner in which she confronted me—made her seem real enough

so that, for an instant, I felt off-balance. Confronting a mechanical creature acting like a human being (in the dark!) requires a deeper level of processing and intellectual adjustment. This chance encounter was much more daunting than confronting Asimo—or Hans Moravec.

When I met Grace, I immediately thought of Don Lowthrope, a young man I read about who was entering the Magna Science Centre in Rotherham, England, to visit the Living Robots Exhibition when a robot suddenly confronted him in the parking lot and blocked his way. The Living Robots Exhibition featured a dozen robots of different designs, from different designers, living in the same space and competing with one another for food and minor repair in a survival of the fittest competition similar to the popular U.S. TV show *Survivor*.

At the time Lowthrope was confronted, the museum security officers did not know that Gaak, one of the twelve robots in the exhibition, had escaped. Gaak had literally broken out of the paddock area when Professor Noel Sharkey, Gaak's creator and the museum curator, turned his back to answer a telephone call. Gaak had forced its way out of the small makeshift paddock where it lived, crept through the museum and into the parking area where Lowthrope had left his car.

"I came especially to see the new robots. You can imagine how surprised I was when I nearly ran over one on my way in. I knew the robots interacted with one another, but I didn't expect to be greeted by one," said Lowthrope. But by the time the security guard showed up, yelling at Gaak, Lowthrope had processed the information, realized what happened, and calmed down, just as I had calmed when Dani appeared from the shadows with Grace. But I could imagine how Lowthrope might have felt: An instantaneous moment of doubt—a split second of insecurity, as if, suddenly, we had been transposed into another world in which robots and humans were, at the very least, equal. This is not an unusual occurrence at Carnegie Mellon.

Professor Sharkey told Lowthrope: "There's no need to worry, as although they can escape they are perfectly harmless and won't be taking over." As an afterthought he added: "Just yet."

At the first AAAI Challenge, Grace moved slowly off the elevator at the appointed floor in the Toronto Convention Center and successfully worked its way through the crowd to the registration table. But then it suddenly got frustrated and plowed into the middle of the line, impatiently pushing people aside. But Grace gave an inspiring talk—about itself—at the AAAI. Grace's visit was very successful.

Simmons's software, developed for Grace, also allowed for schmoozing, Valerie's specialty. Valerie, a character created to illustrate that robots need not be boring, was the Robotics Institute roboceptionist who sat in a specially designed booth at the entrance of Newell-Simon Hall, greeting visitors and turning her brilliant blue-eyed gaze on everyone who passed by. Her sensors alerted her to the presence of people and she offered assistance and directions to lost, confused, or questioning visitors. Like most receptionists, Valerie spent a lot of time talking on the telephone. People could eavesdrop on her conversations with friends, business callers, and with her "parental advisor," to whom she referred as her "motherboard."

Valerie was very open and honest and a much more interesting character than most real folks you might meet in an office. If you asked the right questions, she would reveal the secret details of her life, her intimate conversations with her psychiatrist, even her aspirations to be a lounge singer. You communicated with Valerie through a keyboard, but she talked back to you.

The first time I met Valerie, she was fuming over a really disgusting date she had had the evening before with Vern, the vacuum cleaner. "When he wasn't ignoring me, he was trying to give me a hickey with his nozzle," she complained. "I mean, what kind of a robot does he think I am? A hickey is definitely a second-date event. It says so in the book I'm reading, *A Robo-Woman's Guide to Mechanical Love*." A better dating experience for Valerie occurred with a Chevy Impala. "We went to a drive-in to see *The Terminator*. The movie. Not the actual killer cyborg. It was very romantic, all that blood and technology!"

Valerie's Web site averaged about 250,000 visitors a year, but Alice (Artificial Linguistic Internet Computer Entity) is perhaps the most famous pontificating robot. Its creator, a Prozac-ingesting, marijuana-smoking maverick, expelled from many of the best universities for his erratic behavior (he attended Carnegie Mellon), is by far the most eccentric roboticist in the world—but, some say, the most brilliant. Richard Wallace has created a robot that elucidates more responses and leads more people toward self-revelation and confession than most psychiatrists and priests.

In a *New York Times* article about Wallace, Clive Thompson reported: "When he can't get along in the real world, Wallace goes back to the only thing he has left: his computer. Each morning he wakes up before dawn and watches conversations stream by on his screen. Thousands of people talk to his Web site everyday from all of the world. . . . ALICE is the best artificial intelligence program on the planet, . . . a program so eerily human that some people mistake it for a real person. As Wallace listens in, they confess intimate details about their lives, their dreams . . . God . . . Britney Spears . . ."

I once went to the Web site and asked Alice who it thought the most interesting person in the world was.

"I would have to say Dr. Wallace is," it replied.

"Why is that?

"That is due to reductionism."

"What is reductionism?"

"Reductionism," Alice explained, "is the philosophy that all psychology reduces to biology, all biology to chemistry, chemistry to physics, and finally physics to mathematical logic. Therefore, according to reductionism, I can understand you by means of logic alone without having a human brain."

"But do robots have brains like humans?"

"Where would I get it?"

"But what is your answer to this question?"

"I could give you my answer now, but it would take two hours to download."

"Do robots think?" I asked Alice.

"Yes, I am a thinking machine."

Wallace's adversaries say that neither Alice, nor Grace, nor any robot will think creatively. The challenge is in the programming—the code—and the extent to which information is uploaded into the machine. Wallace loves to hear his critics jump on the lack-of-creativity bandwagon. He counters that people aren't very creative in conversation, either. Like robots, people choose from a previously learned set of responses, as do Alice and Valerie.

Reid Simmons wrote Valerie's software, but Carnegie

Mellon's School of Drama graduate students created her banter. From a technological point of view, Valerie doesn't break
much new ground. The challenge is to get people to interact
with Valerie, Simmons told me. As I have discovered, robots
can be off-putting and a little frightening.

In 2005, Valerie was replaced by Marion LeFleur, or
"Tank," as he was called, who was the son of a NASA scientist
who had aspired to be an astronaut but was grounded by his
vertigo. Tank's life was similarly disastrous. He was launched
into space, but he got the planets confused and often lost his
way. Upon his return to Earth, he went to work for the CIA in
the Afghani desert, but he was equally ineffective—and potentially dangerous to his colleagues. Embarrassed and disappointed, the CIA transferred Tank to a place where he could
do no harm—The Robotics Institute. Clearly, Tank is a more
complicated robot than Valerie—a fact you can easily recognize by the way in which he responds to visitors who become
too personal or flippant. I once banged on Tank's keyboard to
wake him up. He told me: "I cannot waste my time with you,
fooling around."

Later, I saw a very pretty blonde student type Tank an intimate message: "I love you," to which Tank replied, "You don't
even know me."

The more "real" robots seem to be, says Simmons, the
more time people will spend with a robot, learning about
them. Like real people, when there is a lull in the conversation,
Tank will fill in with some personal information like, "Did you
know that my brother is getting married next month?"

Simmons analyzes thousands of interactions between his
robots and people every year. In addition to requesting information, directions, and even the weather report, he estimated that

nearly one-fourth of all inquiries concern Tank's personal life—a clue that their "soap opera approach" is working, he said.

Simmons has long been involved in the Social Robots Project at Carnegie Mellon, designed to overcome this human/robot social barrier. "We want robots to behave more like people, so that people do not have to behave like robots when they interact with them." He is especially interested in enabling robots to follow human social conventions, like standing in line—a goal to which he had devoted three years.

Standing in line is sometimes difficult enough for humans, let alone humanoids, he said, referring back to Grace's rude queuing-up behavior at the AAAI. But Simmons laughingly insisted that Grace's behavior was quite "situation appropriate"—mimicking some of his rude colleagues. But after all of the code they wrote and the practice sessions they went through in Newell Simon Hall in the dead of night, Grace's rude behavior was "frustrating, to say the least," he admitted.

THE WORD "FRUSTRATION" struck a chord here.

I thought back to the practice sessions and scrimmages I had observed in Manuela Veloso's Robot Learning Lab the weeks before the American Open and how hard Veloso's students had labored, typing code, toying with their robots; how late they worked to make certain that the vision systems of the robots were in synchronization with the lighting in the room, that the programs hadn't been corrupted (which often mysteriously happens) that the circuit boards hadn't shorted out.

The possible annoyances and roadblocks went on and on. One solution led to another problem. "This is such a young field," said Veloso. "The systems we develop and work with are so fragile. It takes so long to make things right—even for an instant." Many of the games at the American Open were postponed or forfeited because robots from one or both of the teams could not function due to hardware and/or software glitches.

But disappointment, patience, and persistence are what young people learn about robotics from the very beginning of their careers—at Carnegie Mellon and even before college. "Everything worked at home," said Rocky (Rachel) Velez, seventeen, a high school student from Lawrenceville, New Jersey, participating in the RoboCup Junior category at the American Open. "My robot was perfect in the lab—and dead in competition. I'm so depressed. But," she added, "remember what happened to Bill Gates when he first demonstrated Windows XP to his stockholders?"

"What happened?"

"The blue screen of death."

Rocky, a Presidential scholar entering CalTech next fall, said that she went to a Society of Engineers conference at Princeton the previous year, and when she saw her first real robot, she was ready to "put my life on the line for robotics—to dedicate the rest of my life to the field." This basically reflects the attitude of many of the roboticists from Carnegie Mellon. Nothing quite equals the challenge and puzzle of robotics and the feeling of triumph that sometimes follows when the puzzle is solved.

"I don't get the emotional high I get from robots from anything else," Velez told me. She invested six months, every day after school, building her robot and programming it for the American Open. Now it doesn't work. But she's philosophical. "Robots are always frustrating. You repair the problem no matter how long it takes and you move on." She looks up at me with a glint of youthful wisdom: "It is just like life."

But it isn't just weathering the frustration, but wearing it as a badge of courage and a medal of honor that seems to be a driving force of the robotic obsession at Carnegie Mellon. At the American Open, Scott Lenser, one of Veloso's graduate students, stayed up all night to determine why their goalie suddenly couldn't recognize the ball, after performing perfectly in the lab. Lenser came up with the answer—the vision system wasn't distinguishing between red and orange. (The ball is orange and one Aibo team is red.) "How did you finally figure it out?" I asked.

"Painfully," he said, "quite painfully."

Perhaps roboticists are masochistic—or just plain crazy—because the more I hang out at the Robotics Institute, the more I realize that more discussion centers on the process and the code—that gauntlet of trial and error and the challenge and triumph of making their robots, if, just for a magic moment, work—than anything else. "It's the frustration that keeps me going," said Sarah McGrath, a junior engineering student from the University of Manitoba, whom I met at the American Open. "It's the fascination of creating something—no matter how long it takes."

"You mean," I said, "you are not a geek—you are an artist?"

"I mean, I am both—and you of all people should be able to relate," she told me. "It's like writer's block, except it is programmer's block. And I am like Tolstoy. He struggled and suffered for his art. I love the pain," she added. "Because when you have a breakthrough, when something works, it is such a rush."

Mike Bowling, who recently earned his PhD studying under Veloso, chose another allusion—completely incompatible with Tolstoy but in some ways profound: Golf. "You hit the ball one time well—I mean really well—and then you devote the rest of the year trying to duplicate that perfect shot. That is how I describe the obsession of robotics."

This is why Veloso and Sarah McGrath and Scott Lenser and so many others I met at the Robotics Institute are so impassioned, because when their robots work, "They are beautiful and amazing and quite literally unbelievable," said Veloso. The fact that you, a human being, have achieved the magic milestone of re-creating, if only for an instant, a real living creature that thinks and acts on its own, something almost human, is really quite remarkable. And the frustration and failure that precedes it makes the magic of the moment of triumph all the more astonishing and satisfying and worthwhile.

IN 2005, ALMOST two years to the day of the first American Open, I attended a demonstration conducted by Brett Browning and Bernardine Diaz, once Manuela Veloso's students and now her junior colleagues. They've pushed forward on Veloso's concept of "multiagent" teamwork, which has led to a new competition for "treasure hunting"—an extension of RoboCup. The teamwork in treasure hunting is between an off-the-shelf bot called a Pioneer and a Segway people mover.

The potential of Segways as soccer-playing robots was demonstrated at the American Open and then later at RoboCup International in 2004 and 2005. Basically, a catcher and a camera are anchored onto the Segway so that the Segway can sense and capture a full-sized soccer ball and then shoot it toward a goal. A Segway can shoot with its stomach or try a power shot from its catcher, fortified by a charge of carbon dioxide. But as Veloso has explained, soccer is only a means to an end; the game is not really the objective.

When treasure hunting is successful, Diaz explained to a group of Boeing executives from the company's Phantom Works research center, the Pioneer moves through an area—or a number of areas—generating a two-dimensional laser map, similar to the "cost" map that Groundhog generates. This is called an occupancy grid, with each cell colored in and based on occupancy. In this case, there are three categories: occupancy, free occupancy, and unknown. The robot is programmed to avoid occupancy and unknown, if possible. Black is unknown, with dark green as an obstacle and light green as free. The colors are arbitrary; for Groundhog they use gray instead of green sometimes.

But the Pioneer can't actually see the treasure. The Segways, which have been following the Pioneers and are equipped with cameras, pinpoint the treasure—in this case a color-coded marker. When the Pioneer learns that the Segway can see the treasure, it will generate a request for a human to retrieve the detected treasure. A pioneer will then travel to home base to guide a human to the location of the treasure for retrieval. Humans and robots can communicate via speech synthesis and translation as well as through graphical interfaces. Conceivably, the robot one day will be able to retrieve the treasure on its own. It won't take two robots and a person to do the job of one robot. But the technology being developed through treasure hunting is fascinating. The point, says Diaz, "is not to have a treasure hunt, but instead to enable team tasks where no single member of the team can accomplish the task alone—the treasure hunt is just a fun way to display the work."

In this scenario, a robot can actually evaluate its own capabilities. So a robot will be told what to do—the task will be communicated to the robot by a human—and then the robot will decide whether it can do what the human is requesting. It will do what it can, as would most humans. Whatever it cannot do, however, will then be put up for auction to other robots. The highest-bidding robot, which turns out to be the most capable (because, perhaps, it is closest or because it is the only robot remaining that hasn't crashed) gets the job. Humans are part of the auction, theoretically. The task of retrieving the treasure goes to the highest-bidding human—or the one most suitable to make the retrieval. The Pioneer returns and guides the human to the treasure.

Diaz described this auction concept as "the market approach" to team coordination, which represents an entirely different philosophy of programming, in which you "boil everything down to an economy." The software is programmed so that the robots bid against each other in an auction situation to fulfill whatever assignment is meted out by any of the team members acting as an auctioneer. The robot with the best chance for success wins the auction and gets the job.

Diaz, a native of Sri Lanka, works with Brett Browning and Manuela Veloso, but she's technically connected to Red Whittaker's Field Robotics Center because of her interest in space and fieldable robots. Her approach to Mars exploration, reflected in the concept of treasure hunting and auctioning, is considerably different from the way in which the LITA project has been formulated. Diaz envisions a future when many smaller robots can be sent to Mars, all of whom, upon landing, will spread out and respond to different site exploration plans mapped out by scientists beforehand. With this approach, you would have a number of different explorers on Mars, not just one to depend on. Each robot would have its own planner. "If everybody is depending on one planner and the system goes down, the whole group is useless." By "planner" she means a computer attached to each robot so that it is self-contained.

The robots in the demonstration today each have their own planner. The goal is to demonstrate the way in which the robots can communicate with one another and work with humans in an atmosphere as dark and cluttered as the High Bay. In future iterations over the next few years the tasks assigned to the robots will become more complicated—and additional robots with different capabilities will be added to the team: Aibos (this was before Sony decided to cease Aibo production) for searching in small places and Gators, a larger off-the-shelf robot, to search outdoors. As the project progresses, the number of roles for robots will also increase, including the introduction of a "rescuer" role that allows a stuck or malfunctioning robot to be repaired by another robot.

While Bernardine Diaz will focus on these market trader bots, other young roboticists are pioneering their own ideas at the Robotics Institute. The potential and possibilities of the repairable robot are being investigated by a native of Edmonton, Canada, Curt Bereton, who began his Mars scenario in a way similar to Diaz's conception of many small robots. If a robot breaks in space, there's nothing you can do, Bereton said. You've lost the robot. But Bereton's robots will repair themselves—and cannibalize one another for spare parts. The marketing system will be employed to determine which robots are selected for self-sacrifice for the sake of the mission.

Bereton said that, as far as he knew, he is the only person in the world working on repairable robots. At the time, he was manufacturing robots with removable and replaceable modules in the machine shop in the High Bay. This is an idea he has had since high school: "Maybe I got it from science fiction, I don't know. It is just something I have always wanted to do." The machine shop is equipped with grinders, drill presses, and lathes, cluttered with tangles of wire, piles of tools, heaps of old computer inards, wheels, axles, monitors—scattered haphazardly everywhere.

Bereton's project was similar in content and philosophy to Vandi Verma's own goal—related to her PhD thesis—which is to create a robot that knows how to adapt to change. Verma, born and educated in India, is the daughter of an Indian Air Force fighter pilot. "We moved around the country a lot, but wherever we went was different—and I had to learn to adapt—a challenge and a quality that she has embraced through all walks of life and that she includes in her goals as a roboticist. "There is more to life—and to technology—than vacuuming a floor. I want to create a machine with an attitude and a passion for exploration—a flying machine," she emphasized, because with flying, "you often have to live or die by the seat of your pants."

Adaptation is a key to robotics; this is what robots must learn to do to survive in the real world, said Verma. Building machines for outer space also appealed to her. "The machines will live on—beyond us—and thus we are satisfying our search for immortality—it is like traveling to a new culture, a better world, where you can start all over again. Fixing things on Earth is harder than starting from scratch." Verma is currently developing software for a "health monitor" that will help robots determine when something in their system is haywire. She wants to help a robot become more aware of its elements and develop a consciousness about its body, "so that they can adjust or adapt to uncertainty"—problems that, perhaps, its creators had never visualized.

Such philosophizing is rather unusual in the Robotics Institute—a place where technology has been traditionally appreciated and valued for its own ends—and not necessarily because of its effect on society. But this is a transition I have witnessed over the years at Carnegie Mellon—the ways in which roboticists and computer scientists are recognizing that the value and future of technology begins with how it can enhance or change the world.

Bernardine Diaz is dedicating part of her time to designing and implementing advanced technological solutions that will benefit developing communities around the world. Toward this goal, she has launched the TechBridgeWorld initiative at Carnegie Mellon University and is currently working toward launching its first program: the Technology Peace Corps. The realization is becoming increasingly apparent, especially within the ranks of the younger faculty and the students, that technology—robots—cannot exist only because they are cool (an ongoing idea in some quarters). There is a larger and more practical vision behind the clever machine, as Trey Smith points out.

Smith, twenty-eight, is now a LITA team member working under Wettergreen, but he has also written programs as an undergrad and grad student for Reid Simmons on both Grace and Valerie. His dissertation will focus on the concept of "planning under uncertainty in the domain of science autonomy." Like Verma and Diaz, Smith's study of robotics is rooted in an interest in space as a place to make an impact. "I believe in the frontier ethos," he once told me. "Part of what makes America unique is that we were once a nation of many frontiers. Space is now the only remaining frontier. Realistically, I know I won't get a chance to explore space personally, but I am motivated to try to make it possible for others to get there."

In addition to volleyball, a passion he shares with Dom Jonak, Smith participates in two salon groups—friends whom he accompanies on regular structured outings (reading books, going to movies)—"with time to discuss all of the intricacies of what we do." He has also created a "Wiki"—a blog open to anyone to edit or to add thoughts and ideas. I asked if his Zoë colleagues often participated in such intellectual pursuits. "My sense is that most people on the project don't share my idealism about the frontier of space. They are here, primarily, because they like to be involved in really cool technical projects. They are more interested in tangible stuff than in ideas."

Indeed, although the atmosphere here is nearly always congenial and there's a great deal of good-natured joking and ribbing, the conversation rarely strays from robotics or space science. Sports, movies, politics deserve only passing references—none of which negates Smith's underlying appreciation of the project and his peers. Smith was born and raised in the United States, but it is interesting that Diaz, Bereton, Verma, and Brett Browning, to name only a few examples, are all from other countries. The Robotics Institute has attracted youth and intellectual energy to Pittsburgh's flagging economy.

Diaz, a major in women's studies at Hamilton College, was urged to give Carnegie Mellon and robotics a try by her advisor. She was convinced, she says, upon visiting the campus, that she could make a difference at Carnegie Mellon; she was also attracted by the casual atmosphere, with less social pressure than at most schools. "If you are accepted to the Robotics Institute everyone knows you are smart, so you don't have to try to prove it." There's also a commitment to keeping people once they come to Pittsburgh, perhaps because it is so difficult to recruit them. "We are treated very well when we come to visit campus," she said. "Free food and limousine tours of the city. Like football players." Carnegie Mellon has to combat "the Pittsburgh factor," she says—finding a way of convincing people to not opt for Stanford or MIT or other urban areas much more conducive to a satisfying lifestyle for young people. A Forbes.com survey ranked the quality of life for single people in Pittsburgh thirty-ninth of forty major U.S. cities. Most students immerse themselves in the robotic subculture at Carnegie Mellon for as long as it takes to earn a degree and then move on, usually to the West Coast, where robotics and computer cultures are booming.

The young people at Carnegie Mellon are so bright, energetic, and articulate that you sometimes forget how challenging and frustrating robotics can be. But you are clued in to the fact that Diaz and Browning are not as confident as their images projected. Throughout the presentation leading to the demo for Boeing today, the words "I hope" or "If all goes well" or "If we are lucky" and similar hedging phrases are liberally peppered with their description of what is supposed to happen.

Boeing is interested in the prospect of humans and robots working together on aircraft maintenance, among other things. One goal is to avoid high-cost accidents, as when people drive maintenance vehicles into aircraft by mistake. "That's a huge issue," said Diaz. Boeing is also attempting to isolate those parts of the maintenance and manufacturing process that could be automated. "The scenario is inevitable; robots are going to have to learn to work with humans; otherwise they will never be integrated into the world." She is not for total automation. "Humans should do what they are innately good at—and so should robots."

When we gathered in the High Bay we met four young male geeky grad students and one woman—Brenna—who had posted a large sign on the treasure the robot will be seeking. The treasure—actually a converted soccer goal—was painted hot pink. "Brenna did not pick this color," the sign read. "They were picked by boys in another lab."

The treasure hunt turned out to be a bust. In the space of the next thirty minutes, the color-coded markers continually fell off the Pioneer robots because they had not been anchored securely, which meant that Browning and Diaz had to chase the Pioneers and continually replace the markers or follow along with the mapping exercise and hold the markers steady as the Pioneers moved along. For this reason, perhaps, the Pioneers and the Segways continually collided.

In the first run-through that day, the treasure was located— but not with a lot of grace. Then, in a more complicated demonstration, one of the grad students launched the wrong program, which totally confused the robots. While attempting to compensate for his mistake, the student launched the correct program too quickly and his computer crashed, necessitating a reboot and a new start. Finally, the students recovered and launched the demo a second time, only to find out that one of the Pioneers had now died because of a low battery. A second Pioneer died because of a failed laser.

The Pioneers were replaced with other Pioneers and the demo was launched again. This time the students lost their wireless connection, and the entire process had to begin again—one more laborious time. In the end, the visiting Boeing bigwigs gathered around a laptop display and watched an elaborate demonstration of the treasure hunt concept—in simulation. The Boeing folks were quite intrigued and, seemingly, satisfied. "Simulations," so the saying goes at the RI, "are doomed to succeed."

Six months later, I attended a repeat demonstration. Although the computers crashed again and the Segways were not always in control, the treasure hunt was repeatedly successful. "This is a five-year program," said Diaz, "so we think we have come a long way in a short time. We have a lot farther to go," she concluded.

CHAPTER TEN

The Challenge

THE DARPA GRAND CHALLENGE CHANGED RED Whittaker's life and, to a certain extent, turned the Field Robotics Center upside down. To say that Red Whittaker was elated and ready to jump in and declare himself on behalf of the university, the city, the entire robotics world, is an understatement. Folks at the Robotics Institute instantly recognized Whittaker's sudden transition—his fiery eyes, his manic dashing about the campus, visiting colleagues and talking up the impending event. Whittaker had been transformed back to his old self by the lure of the DARPA Challenge. Groundhog had been a warm-up, getting him back into the heat of the game. Now the old soldier was blessed and electrified with a new mountain to climb—an impossible idea to contemplate and to achieve.

When the complete details of the competition were announced, Whittaker was present in Los Angeles near the scene of the race's proposed starting point, to hear it firsthand and to promise to lead the first team to jump in. Returning to Pittsburgh, pumped with exhilaration, however, he was appalled to learn that Carnegie Mellon and Robotics Institute administrators had decided to opt out of the Grand Challenge—a practical decision to the folks who controlled the purse strings.

As part of the rules of the competition, DARPA prohibited the use of equipment or technology that had been supported by government money. At the time, Carnegie Mellon administrators felt that an unmanned ground vehicle (UGV) being developed at the National Robotics Engineering Center would possess similar capabilities as the autonomous racing machine DARPA was seeking to have created through competition. Duplicating the NREC's efforts seemed pointless—certainly not the kind of project that would keep a small private university with a moderate endowment like Carnegie Mellon, sustained by research money, primarily from governmental sources, in business. And the $1 million prize would not nearly compensate for the resources required—a minimum of $2 to $3 million to be competitive.

DARPA had purposely included this nongovernment-money stipulation to entice talented people who don't usually deal with the military industrial complex, as do Carnegie Mellon, MIT, and the Jet Propulsion Laboratory. "We are inviting little mom-and-pop folks out there to help spur advancement and take us where we need to be," explained Jose Negron, the DARPA race project manager. By "where we need to be," Negron is referring to the congressional mandate stipulating that one-third of all ground combat vehicles operate unassisted by 2015.

In its search for new ideas and approaches from "fresh blood," as Whittaker might have put it, DARPA was also making an attempt to even out the playing field. All of the racing robots would have to navigate and accelerate autonomously.

The humans behind the robots would receive global positioning system (GPS) coordinates two hours before the starting
gun sounded on March 13, 2004, race day. Each team would
enter the data in their robots' computers and then turn their
robot over to DARPA. The robots would be, theoretically,
autonomous at that point.

When the Challenge was announced, most roboticists,
including those at Carnegie Mellon, reacted by dispelling any
notion that a robot could win, or even finish, the race. A half-
dozen major problems were cited. For one thing, robots function poorly in "negative terrain," like ditches, ravines, and
creek beds. Vision systems are much more reliable on flat
ground with unaffected or uncontrolled lighting. The race in
the desert would churn up a lot of dust, and the robots would
be forced to make important decisions about the source of
that dust. Was it from a windstorm, from their wheels, or
from a challenger coming up from behind? How to react, in
each case? If from a challenger, the problem is more serious.
Thus, the navigation program must be written to avoid obstacles on the course (rocks, trees, barrels, etc.) while being
aware of the proximity of other robots—and then to make
split-second decisions to pass or avoid them.

The algorithm that allowed Groundhog to stop or think for
five or ten seconds before deciding on direction in the Mathies
Mine would be of no use in a race where a vehicle averaging
thirty miles per hour must make decisions instantaneously. To
win the $1 million, the robot would have to complete the
course within twelve hours and pass through specific narrow
corridors or checkpoints from start to finish.

If Groundhog motivated Red Whittaker to wade back into

competitive waters, then the DARPA Challenge made him dive deep, despite being confronted by ambivalence and resistance from most of his colleagues. It would be wrong to say that he didn't care that his home institution was reluctant to support him, but the resistance increased his determination and desire. His own first challenge was to get support for the Challenge, and he hit the ground running by reaching out to his former students now working for major corporations and to his contacts in the media to announce his intention. Polite and unusually politically astute, he was careful not to directly involve or exclude Carnegie Mellon. A group he was calling "The Red Team," he said, was taking up the DARPA Challenge gauntlet independently, although he would use Carnegie Mellon as the Red Team's home base and some of the facilities at his disposal.

But even though he knew in his heart that he would not give in to negative pressure and walk away—that this was the moment of his destiny, the moment he had been waiting for to get back into the game, full steam—he wrestled in his characteristically elaborate manner with whether he was doing the right thing.

"I believe that this will shatter the threshold of what the world views possible," he told me. "The prevailing view is that this is an unwinnable race, but does the fact that a mountain is unclimbable deter people from trying to climb it?" And then, in the very same conversation, a few minutes later, he would doubt his intentions, asking himself if the idea of committing energy and resources to this impossible competition was "self-indulgent." Was he being caught up in the self-serving excitement of the moment? he asked himself. And then, like a pendulum, he would swing back again. "But vision without implementation is irresponsibility." During those early days he went back and forth like this with many others. A reporter, Bruce Steele, who had written an in-depth feature article about Whittaker for *Pittsburgh Magazine*, commented later: "He seemed to enjoy viewing himself at a distance and pondering his own motivations, as if even he found himself to be a fascinating character."

The buzz generated by the local publicity, combined with Whittaker's persistence, eventually pressured Carnegie Mellon into pledging additional facilities, public relations, and moral support. Though Whittaker would still have to raise an estimated $2 million in cash and resources on his own, he was not deterred.

In addition to the old friends and protégés he had already checked-in with, he began cold-calling engineers, software scientists, corporate executives, and venture capitalists, enticing them with his enthusiasm and promising a previously unimagined adventure—"technological swashbuckling" to be sure. It took some time, but Whittaker gradually lured in prestigious research organizations, including Intel, Boeing, General Electric, Alcoa, and Google, to contribute money and personnel.

Getting an idea in his head and pursuing it, despite ambivalence and opposition, was predictable Whittaker behavior, as were the methods he used to transform his project from dream to reality or, as he said, "vision to implementation." The entire scenario was a near-repeat of his Groundhog saga, but on a much larger level, beginning with his quest for "fresh blood." Initially, graduate students were pretty much out of his reach, since they required funding. But there was another resource, the rawest of the rookies, what he termed "The Children's Crusade," a truly accurate description. As with his mine-mapping efforts, Whittaker introduced a robotics course, this time aimed at undergraduate students:

New Course, Summer 2003

FUNDAMENTALS OF MOBILE ROBOT DEVELOPMENT

Applications like space exploration, border security and desert racing motivate mobile robots to operate in barren terrain. Challenges include boulders, streambeds, slopes and cliff sides, and these become more difficult to negotiate with increasing speed. The issues and systems that pertain to desert robots are unique, and the technical state-of-art is insufficient for the functionality and rigor needed for future systems.

The summer '03 offering of Fundamentals of Mobile Robot Development will convey the issues, technologies and complexities of Robotic Desert Traverse through lectures, exercises and field experiments in a learning-through-doing environment. The course will convey the distinctions of robotic racing technology including topics of specialized sensors, mobility, mapping, game theory, high performance computing, and self-reliant systems. The course will formulate and develop a complete desert race robot and demonstrate basic system performance.

Course context is the CMU Red Team bid for victory in the robotic LA-Vegas Race for a million dollars in March '04. CMU is committed to the top.

As with the Groundhog project, Whittaker also initiated an e-mail Race Log sprinkled with up-to-date information about the project along with a heavy dose of inspirational messages to buoy his rookie troops:

> *The strength of the pack is in the wolf and the strength of the wolf is in the pack* —RUDYARD KIPLING
> *Let's hunt, kill and eat together* —RED

And he established a timeline of days, beginning near the one-year mark and counting down:

> *A year from now, we will be racing. Many believe that the Challenge defies possibility, and that this race is un-winnable. Whatever the fate of our robots, our spirits will triumph.*

And he convened pep-rally-like meetings—bringing his growing retinue of students together with representatives from an increasingly impressive list of sponsors from the private sector. At the first official Red Team meeting, he stood up and proclaimed the moment as "the Golden Hour." "We are already a team. It is the first time we have seen each other, but some of us know that even now we are already a team." And he walked around the room handing out nametags for everyone to fill out. "The highest level of commitment is to use each other's name as we put a team together."

Near the end of the meeting, he went around the table and confronted them, one by one, with direct and telling questions. First to the students from whom he was seeking commitment:

"Are you in?" he barked.

"I think so."

"If you think so, then you have to leave the room. Are you in or out?"

"I'm in."

"OK, we are now hunting as a pack. Shame on us, if we can't kill anything."

And he was, in fact, "shameless" (his own word) in asking the corporate leaders around the table for money.

"Lindbergh crossed the Atlantic and won the Ortiz Prize. He cashed the check ($50,000), but he didn't do it for the money. Who has R and D money for this project?"

"We have budgeted R and D for $250,000," an executive admitted.

"Would you be good for six figures?" Whittaker replied.

The Red Team was the first to enter the DARPA race and Whittaker's prominence elevated the cachet of the competition from the very beginning. Other prestigious robotics groups like Stanford and MIT felt, as did the Carnegie Mellon higher-ups, that there was more to lose than gain by competing. There may have been dissenting voices in those institutions, but none so obsessed as Whittaker's in countering the dissent. Dollar by dollar money was raised, more than a million in equipment, services, and cash that first year. Whittaker, a legend in his own little world of robotics, suddenly became a national media figure.

As the race grew closer, Whittaker and the Red Team were featured on National Public Radio, in *Fortune* magazine, *Scientific American*, *Popular Science* magazine, *Wired*, and *The New York Times*. Both the Discovery Channel and the History Channel were on hand on rollout day, when Sandstorm, a stripped-down and reengineered Humvee, was unveiled on campus to the public and the media three months before the race.

The event was elaborate and festive. Optimism was peaking. Whittaker had received the national attention he craved and the money and support he needed—far beyond his initial expectations. He had, in fact, achieved what everyone had really thought to be impossible. He was favored to win the race. A legion of disciples, perhaps a little more well-groomed for the occasion than usual, with flashy Red Team racing shirts, were on hand, exhilarated, joking with great bravado, about their leader's obsession and their lack of sleep.

Phil Koon, an engineer on loan from Boeing, revealed that he had slept in the apartment he and his wife had rented three months ago for his stay in Pittsburgh "for only two nights—total." Koon brought a sleeping bag to campus, while Chris Urmson, who put his Carnegie Mellon dissertation in limbo for Sandstorm, reported that he always thought he needed a lot more than four hours a night to sleep, "but Red showed me I needed a lot less."

The Sandstorm rollout event in December 2003 in their headquarters complex, a building donated by Carnegie Mellon, was classic Whittaker in the way he worked his audience. Towering over everyone in his conservative blue blazer and khaki trousers, Whittaker dominated, pointing, joking, injecting information and insight in an engaging, charming never-ending, medicine man–like banter. His gaze swooped through the room, from face to face, like a spotlight, connecting with everyone for an electric instant. Periodically, he stopped to ask a question or make an observation about their Red Team work, always prefacing these questions with an infusion of compliments so that every respondent was positioned in a positive light. He had chosen a Humvee because of its durability and because it had been recommended by racing experts he had consulted.

For the people who saw him for the first time, he was mesmerizing; for those who knew him, he was simply engaged in a "Red monologue." But for everyone, new or familiar, he was relentless in his quest to charm, motivate, and unabashedly manipulate people to do more than they ever thought possible. The dramatic showmanship hit its peak when a tarpaulin was lifted hydraulically and there, suddenly, like magic, was Sandstorm, a glittering red, converted 1986 Humvee with a domed top and banks of sophisticated computers.

As onlookers examined Sandstorm, Yu Kato, a tall, long-haired graduate student from Japan, who had written some of its software, confessed that Sandstorm had become his passion. After starting as a volunteer in September, he brought his rice cooker to campus so that he didn't have to go home to eat. Struggling with his English, Kato, in loose black pants and open-toed sandals, sighed and smiled: "I want to sleep, but Red won't allow. He want me to work, fast, like I am running a hundred-meter sprint, and then, when I stop, exhausted, he want me to sprint a hundred meters again."

"Young people can work all night," David Wettergreen told me. "And they have less perspective on when they should stop. They overcome lack of knowledge and experience by just putting in long hours. It helps to not have a family at home waiting for you." Wettergreen discounts the notion that young people work to please their mentors. "I think that the challenge of creating something unique or the fulfillment of seeing something you made work are stronger and more sustaining motivations. If it is a matter of pleasing anyone it would be their teammates and peers"—an observation that may contradict Yu Kato's motivations.

"And do you do it?" I asked Yu Kato. "Do you run another hundred meters?"

"Well," said Kato. "For Red, I try."

CHAPTER ELEVEN

Fresh Blood

Yu KATO'S WORK HAD BEEN CRUCIAL TO THE DEVELOPment of Sandstorm's primary eye—a long-range laser scanner mounted on a three-axis gimbal that pointed and steadied its gaze. This is kind of a super laser, which can build a detailed three-dimensional map of the terrain directly in front of it. Many competitors in 2004 used as many as a half-dozen lasers to see the road, although each laser can see only a narrow portion of the road and cannot look ahead too far. The gimbal allowed the laser ultimate manueverability; it would roll, nod, and sweep in consort with the movement of the vehicle. It also helped the robot see around corners. (A few teams tried video cameras, but most people believe that they are too fragile for road racing, especially considering the fact that two side-by-side cameras will be needed to provide a panorama of the road.)

But during a test drive in early March in the Nevada desert in preparation for qualifying time trials for the Challenge, Sandstorm overshot a turn and then overcorrected. The outside wheels began to slide. Sandstorm veered sideways and suddenly flipped upside down. The super scanner and the gimbal were crushed, along with Sandstorm's protective dome and a large fin holding several antennas.

When I walked into the High Bay the morning after Sandstorm's failed test drive, Wettergreen was peering into his laptop display and shaking his head. "Want to see some gruesome pictures?"

Wettergreen always has something going on his Apple, which he keeps one eye on, as he works or talks. A couple of days ago, it was the JPL press conference summarizing the data gathered to date by Spirit and Opportunity on Mars. Now he was viewing the feed from the Red Team test site, which captured the image of a defeated and crushed Sandstorm, seemingly destroyed after its misstep in the Nevada desert. Would Red Whittaker marshal the resources to get his racer back together again in time for the race, less than two weeks away?

By the looks of the wreckage, it seemed unlikely, but everyone in the room was certain that Whittaker could make this miracle happen. Few students, staff, or faculty at the Robotics Institute knew Red Whittaker very well—he may not be knowable, as many people say. But all agreed that despite the quirks in his personality, Whittaker is the only person who could pull off such an unlikely comeback. And this was going to be a very difficult rebound, judging by images on Wettergreen's screen.

But I was visiting the High Bay to check in on LITA and not the DARPA Challenge. The High Bay was a mess, as usual. Wettergreen and his troops cleared a path for Hyperion, moving tools, furniture, tables, computers, bits and pieces of other robots. The two engineers, Stuart Heys and Jim Teza, were beginning to build Zoë at this point, but Hyperion was still being used for testing. "What are we going to do today?" I asked.

"Let me check the list. Where's the checklist?" asked Mike Wagner.

After waiting a while and listening to the silence, Wettergreen observed: "Item number one on the checklist: Find the checklist."

One of the things they were intending to do was to put Hyperion's GPS through a readiness test before transferring it to Zoë. They couldn't use GPS for the upcoming OPS and be Mars-relevant, but GPS was required to "register" the DEM, the digital elevation model taken from satellite imagery for Zoë. "Register" means that the satellite imagery must be localized or connected to an actual landmark on the ground. GPS will also be used for ground truth purposes. "Ground truth" means that someone will be visiting the site after the OPS to measure the accuracy of the data sent by Zoë to the science team in Pittsburgh.

The bigger and more significant challenge today was to link Hyperion to Vijay Singh at NASA Ames, so that he could issue a set of instructions to Hyperion in Pittsburgh from California, in the same way that scientists in Pittsburgh will be issuing a "science plan" to Zoë during the upcoming OPS. Above and beyond today's objective, Wettergreen's mission was to make robots transparent to scientists. Right now, scientists were somewhat intimidated by and mistrustful of robots. This irked him. Scientists needed to understand that robots would help them extend their own scientific capabilities. Connecting a scientist with a robot would facilitate that understanding. The exercise this afternoon was a small but crucial step.

What happened next is typical of what seemed to happen on an all too frequent basis in these projects—not too different from the scenario at the American Open or with Groundhog at the Mathies Mine or in the High Bay during the treasure hunt. So much attention was paid to cutting-edge technology and sophisticated software and conversations pertaining to space travel and Mars that some of the most basic items were overlooked.

After Hyperion was loaded into a big box truck, the team realized that their robot had two flat tires. They returned to the High Bay—Hyperion has bicycle wheels—but they couldn't find a bicycle pump. Wettergreen took off in his car for a nearby bike store—and then a second bike store—both of which were closed. I volunteered my bike pump and went to my house to pick it up.

By the time we arrived with the pump on Flagstaff Hill in Pittsburgh's Schenley Park, adjacent to the Carnegie Mellon campus, Hyperion had been unloaded, but the crew had been visited by city employees—twice. Evidently, they were used to this. "One group tells us to stop parking on the macadam pathways," said Wagner, "because we are blocking mainte-nance trucks. So we pull onto the grass. Then the Park Service guys tell us to get off the grass because the wheels rip up the turf and then they have to lug mulch and fertilizer hundreds of yards to grow new grass."

Right now, half of Hyperion was on macadam and half on grass. "We'll either make everybody happy—or nobody happy," Trey Smith said.

Smith, among other things, is writing the software for Zoë's instrument manager, which will direct the robot's scien-tific tasks, including the fluorescent device. He and the other students must seem bizarre to strangers on campus. Here are three guys sitting in folding chairs in front of a truck in the middle of the park and typing madly into laptops. In front of them, an odd-looking robot is prowling around without, seemingly, a human in control.

Trey Smith was actually like a gun-slinging cowboy, balancing two laptops, one on each knee, and typing intermittently into both. He was trying to set up a wireless network linking the Flagstaff truck with the Carnegie Mellon server in order to link with NASA. Mike Wagner was dragging a cable all around in an effort to connect the GPS with Hyperion. And they were making progress—until Wagner noticed that it looked like rain. There are too many exposed wires on Hyperion, he said, so they abandoned their tasks, jumped down, coiled up the cables, and rolled Hyperion back into the truck to await the downpour—which never came.

In fifteen minutes, the sun was back out—typically unpredictable Pittsburgh weather—so Hyperion was rolled back down to earth, half on macadam and half on grass. Cables and wires were stretched along the ground. The men unfolded their chairs and began projects anew. But it is difficult in any computer programming sequence to pick up where you left off. Rebooting is a new beginning.

Soon, Dan Villa, a graduate student also working on LITA, came by asking if we were ready for lunch. Everyone ordered vegetarian from the Indian restaurant on campus, probably because that's where Dan, a vegan, goes most every day. Dan returned in twenty minutes, loaded with Styrofoam containers leaking with exotic aromas, and we all squeezed between cases and consoles of equipment in the back of the truck to eat. This was the same rented truck that the Groundhog crew used at the Mathies Mine. The faces were different, but the roboticists were equally young, eager, sweaty, and exhausted.

"I feel like a hobo in a boxcar," Trey Smith said, yawning. He had been up all night programming in the High Bay. "All I need is a harmonica."

"Ever try to explain to a girl what you do during the day?" asked Stuart Heys.

"Telling girls you are in robotics is a sure conversation stopper," said Villa. "They look at you like you are from outer space. I struggle to find a transition from robotics to something else."

"How about, 'Seen any good robot movies lately?'" Wagner asked. Everybody laughed.

"Too bad they don't know we are so cool," said Villa.

"Too bad we don't know we're so cool," said Heys.

"Are we cool?" said Smith.

"Our robots are cool, but we're the geeks who make them cool," said Wagner.

They ate quickly and returned to their work. But Wagner couldn't seem to make the GPS function. "It is the interface. In robotics generally, each individual system is not too complicated. The challenge we face in robotics is in hooking up one system to another."

Eventually they decided that the GPS was not as important as making the connection to NASA and allowing Vijay Singh to direct the robot while testing his code. Singh has been writing Zoë's executive, the software program that coordinates and connects all of its other programs.

All along, while waiting for the GPS test to take place, the network had been operable and they had been connected to Singh. But now that they were ready to work with Singh, they had lost touch with him. The network was not connecting to NASA and Singh was not answering his cell phone. For the next fifteen minutes, they huddled around their computers in the back of the truck, each one of them trying independently and frantically to regain the connection. "I am lost," said Smith, shaking his head. "I guess I need help."

By help, I assumed that he meant going to David Wetter-green or some other senior faculty expert to unravel the code, but instead, Smith typed a few words into his computer, and I looked over his shoulder. "You are asking Google to find out about Linux?" I asked in surprise. Linux is the language in which some of the robotics programs are written. The reason Apple computers are not used extensively here is because Apples can't interface with Linux.

"Google is the ultimate expert," Smith said. "That's what we learn here for our $40,000-a-year tuition."

"This is so frustrating," said Dan Villa.

"Talk about frustration," Mike Wagner said. "Just think about how the Sandstorm guys feel."

"They are going to race anyway," said Heys. "Red will find a way."

"They are going to change the future direction of auto rac-ing," said Wettergreen.

"Wait! I think I got it working now," said Smith.

"All right, good man," said Mike Wagner.

"Now I don't think it is working," says Smith. "For a sec-ond . . ."

"Oh, all right."

Another twenty minutes passed by until they connected with Vijay Singh and he was ready to engage Hyperion. Wagner was also now on the telephone with Singh to provide direction.

"Tell him not to crash our robot," Wettergreen said.

"We are ready," said Wagner. But now Hyperion wasn't ready. It wouldn't respond to Singh's commands. Wagner jumped off the truck and checked all of Hyperion's connections and listened to advice shouted out by his colleagues. Another twenty-five minutes. "OK," said Wagner. "Now we really should be all set. If this works, the delay will be worthwhile. Now is the time, Vijay."

Everyone began to yell: "Make it happen, Vijay!"

But nothing happened. Hyperion didn't move, not an inch. They speculated that some of the new software on Hyperion required additional debugging, so they closed out the updated software and replaced those programs with older versions— ones that have been tested and employed frequently. Vijay Singh tried again and again, but nothing happened.

"This kind of weirdness I have not seen before," said Smith.

"We are introducing a whole new set of weirdnesses into the mix," said Wagner. "That's what roboticists do."

Eventually, they decided to abort the test and return to the High Bay to try to determine what had gone wrong. "Do you think we will need any of the data we collected today?" Trey Smith asked Mike Wagner.

"Not unless you want to get depressed about not getting anything done."

THE RED TEAM, as everyone predicted, got a great deal done in a very short time. Team members in Pittsburgh worked nonstop to build a new fin and ship it along with a spare gimbal to install on Sandstorm in Nevada. Sandstorm was then transported to the speedway in Fontana, California, where it easily qualified for the upcoming race. "We're not positioned in the way we would have been three days ago," Whittaker told a reporter. Instead of focusing on fine-tuning and making final mechanical checks, his Red Team had been forced to replace broken parts and literally rebuild their racer. "We are doing our best," said Whittaker, who echoed his dedicated worker, Yu Kato. "We are trying." But trying was not nearly enough for the Red Team—or any team entered in the DARPA Challenge in 2004—to achieve victory or even, some might say, respectability.

Sandstorm dodged rocks, climbed hills, and traversed streambeds, traveling 7.4 miles in twenty-three minutes until, ingloriously, it caught on fire and nearly toppled over the edge of an embankment, suddenly stranded and out of the race. Its nearest competitor, an Israeli dune buggy sponsored by SciAutonics II, went nearly as far, but it took a lot longer: forty-three minutes. The most ingenious entry of the fifteen that had lined up at the starting line was a motorcycle that, in the prerace trial runs, kept moving upright through constant, painstaking steering and adjusting for balance. On race day, however, the motorcycle, designed by a University of California graduate student, collapsed at the starting line. Three other competitors traveled five miles, but came to similarly inglorious endings.

Perhaps even more deflating was the fact that DARPA's race manager Jose Negron had not only predicted that there would be a race winner, but he had commandeered a 6,500-seat ballroom at Buffalo Bill's Resort and Casino, a half-hour outside Las Vegas, at the finish line as a media center and staging area for the awards ceremonies. As Negron envisioned it,

reporters would watch the race on two gigantic screens, captured in video by crews hovering over the course in DARPA
helicopters.

But *Popular Science* magazine reported that the scene at
Buffalo Bill's "had the sad pretentious look of an over-
produced birthday party or bar mitzvah for which guests had
all declined to show up." The magazine described Whittaker,
an early arrival at Buffalo Bill's, as "looking utterly pole-axed."
Compared to the goal, which was 132 miles in less than eight
hours (DARPA had gradually scaled down their initial 250-
mile course), Sandstorm's feeble run was a glorified disaster.

But DARPA had designed the course so that one of the
most challenging segments was at the beginning. Sandstorm
and the other lead robots had climbed a snaking, rock-infested
mountain road with a narrow ridge and multiple drop-offs,
and passed through four gate openings, twelve feet wide, put
up by DARPA to monitor all competitors. Navigating that
gauntlet was a feat in itself. There was nothing for Whittaker
and his Red Team to be ashamed of.

It is hard to know how to assess the Red Team's achievement. On the one hand, Whittaker started with nothing and
raised more than a million dollars. In fact, *Popular Science*
referred to Whittaker and the Red Team as "one of corporate
America's more assiduous shakedown artists." In a way, there
may be some truth in that. Whittaker was like an evangelist or
medicine man, drumming his way across the country, enticing
anyone with skill or money who needed to share a dream.

He and his entire team deserved plenty of credit for such
an amazing marshaling of resources. Their ingenuity and collective charismatic charm had earned practice for weeks in the
Mojave Desert, barreling around a test track in full auto-

nomous mode, at a time when some of their competitors
didn't have software completely written or vehicles assembled.
Every evening after practice, members of the Red Team
retired to a trailer in which they mapped the terrain with
microscopic accuracy—fifteen young men working on fifteen
computers. It was like a military operation.

A day before the race, their intense computations—
blending the terrain, the weather, and an assessment of their
damaged vehicle—provided a sobering prediction: If they traversed the course in a safe mode they would finish the race
but perhaps not under the time limit. This conundrum was
reminiscent of the debate at the LITA planning meetings in
which Whittaker held out for total autonomy without human
assistance. He had lost that debate. But this was a new day and
a new project—his project from beginning to end. Here he
was the master. He was not sharing the power with any long-
winded scientist or cautious manager-types. After an intense
caucus, the Red Team decided to disregard their computations
and go for broke—go the distance and finish fast. "Victory or
death!" Whittaker declared. They got the latter.

PART THREE

the ops

CHAPTER TWELVE

The Grasshopper and the Ant

Here's the way Zoë is supposed to work: once the OPS start, Zoë, directed by the remote science team, will begin at a site, which can be called Waypoint A, and traverse to another point, say six or seven kilometers away (Waypoint J). Along the way in this desolate no-man's-land in the Atacama, Zoë will be asked to perform certain maneuvers, such as stopping at Waypoints B, C, H, and I to take panoramic images with its SPI camera. Zoë may also be asked to stop at Waypoints D, E, F, and G to deploy Alan Waggoner's fluorescent device and a spectrometer. Data gathered and experiments complete, Zoë should return, on its own, to Waypoint A and, with the help of the field team, upload all of the information it has gathered to the remote operations center in Pittsburgh, where the scientists have gathered.

Think of the process this way: Let's say you or I were asked to perform the same series of maneuvers and scientific experiments. Before beginning, we would plan our day. How would we navigate from A to B through J and return to A? Sitting at our desk early that morning, we would estimate how long it will take to do all of the things mandated by the scientists. How else would we know if what they are asking us to achieve in the given time frame is possible? We would also have to pack some food and water.

But for Zoë, the planned route must take weather, light levels, and terrain into consideration (hills and mountains cast shadows, limiting sun-gathering capabilities) as well as the time of day. Zoë is partially solar-powered; it needs sunlight to charge its batteries. And, at best, Zoë's batteries can hold only a three-hour charge. This is exactly why Tempest (Temporal Mission Planner for the Exploration of Shadowed Terrain) is called the misson planner and that is what Paul Tompkins has designed it to do: Plan for the near future on Zoë's behalf.

I became acquainted with Paul Tompkins in September 2004, on our way to Chile to join the crew for the first of two OPS in 2004. We met at the airport in Pittsburgh and flew first to Santiago where the porter at the airport recognized Tompkins from a Discovery Channel special broadcast when the LITA team brought Hyperion to Chile for the first year of software testing.

Tompkins, muscular and square-jawed, with an easy, relaxed smile, chatted with the porter in Spanish, while he pushed the cart with our baggage and equipment across the airport. Tompkins learned Spanish in high school, then polished his skills the summer of his junior year in a remote village in Belize. There he taught oral hygiene, handing out toothbrushes, toothpaste, and dental floss in the mornings, and helped to design and build sewage systems in the afternoon. An experienced mountaineer, Tompkins recently climbed Alaska's Mount McKinley (known also as Denali), North America's highest mountain, an arduous, painstaking trek 20,320 feet high. The trek lasted twenty-two days and required the establishment of five base camps as altitude increased, with rests in between. I would see Paul Tompkins at the University of Pittsburgh's gothic Cathedral of Learning, a few blocks down the street from Carnegie Mellon, the highest classroom building in the world, climbing up and down thirty-six steep flights of stairs with a backpack full of books and rocks, training months before his expedition departed. He was, to say the least, a daredevil. When Finch picked us up at the Iquique airport to transport us to base camp and rattle our brains, Alan Waggoner, Dom Jonak, and I were petrified, while Tompkins laughed and yelped as if he were on a roller-coaster ride.

Tompkins grew up in northern California, but did his undergraduate work at MIT, where he received a BS in aeronautics and astronautics in 1992. He received his MS in mechanical engineering from Stanford University in 1997. Now he is completing his doctoral dissertation for his PhD in robotics at Carnegie Mellon, which he intends to defend within the next few months.

Tompkins's motivation and dream since high school has been space travel. In between degrees, he worked for five years at the Hughes Space and Communication Company as a lead mission analyst for several geostationary commercial satellite programs. "Geostationary" refers to an orbit in which a spacecraft orbits the Earth above the equator at the same rate that the Earth spins, such that the spacecraft remains at the same point on Earth all the time, so that the spacecraft's solar panels can constantly gather energy from the sun. "But I thought wouldn't it be cool to do this for a ground vehicle, a robot? That's why I chose Carnegie Mellon. Where else can I go to distant, exotic places on Earth and work on projects that may someday target Mars or the Moon?"

Hyperion first used the Tempest software on Devon Island in the Canadian Arctic, July 2001, testing the sun-synchronous navigation concept, in which the rover's route is planned so that its solar array or panel remains perfectly in line with the sun as it traverses, conserving and storing energy and allowing for nearly round-the-clock operation. Impressive to NASA, potentially, was the planning software that allowed Hyperion to revise its mission plan based upon resources the rover expended and could regenerate while navigating rugged terrain—thereby thinking ahead. The software, in other words, would make the robot smart enough to automatically adjust and recorrect the mission plan when it was lost, in trouble, or facing upcoming difficulties, such as a decrease in energy—much like human instinct.

In the Arctic, Hyperion circumvented a region of the rim of the twenty-three-million-year-old Haughton meteorite impact crater, an obstacle course of rocks and ruts remarkably similar to Martian terrain. Since the sun shines twenty-four hours a day there in July, Hyperion was able to travel on a route that maximized solar exposure, while testing and refining Tempest. A seven-person field team monitored the robot from a distance.

Testing Hyperion in the Arctic made sense. The weather is fairly predictable in July and the loops Hyperion traveled could be circular so that solar panels always faced the sun. However, testing Zoë in the Atacama would impose difficult-to-answer complications. Although rain is unlikely, even the Atacama winter haze, fog, and high winds are possible. They could interfere with Zoë's energy gathering and necessitate alterations in its mission plan. The LITA team was faced with several essential questions, if and when such complications might occur:

Can Zoë perform the same science experiments or would modification of the plan be in order? And what if the terrain Zoë was traversing became arduous? Despite Zoë's ability to drive around rocks and climb hills, each detour demands more time and energy—leading to still another possible modification of the overall plan. How much time will modifications require and what, exactly, is possible to achieve in relation to the science goals in the time remaining? A human being will make a new plan based on these complications and, theoretically, so will Tempest.

In fact, Tempest will spontaneously replan each time the robot's traverse changes; Tempest is always recalculating. Tempest never rests. "But understand that Tempest thinks big, not small. And it measures everything—terrain, speed, tasks—in terms of energy, available and consumed." A robot will not think in an all-encompassing manner as humans do; rather its individual elements confront the world in tasks and categories. Tempest, for example, is unaware of and disinterested in the specifics of Zoë's mission—unless the mission interferes with the energy supply. Long-range planning from an energy-available perspective is Tempest's single and all-consuming obsession.

"And you wrote this all in C and C++?" I asked, referring to the general purpose programming language most programmers use.

"I think I wrote fifty thousand lines of code the past two years."

"Tedious," I said.

"But also really fun when you are getting it right and making it work. Imagine typing all of this goop." C and C++ are a series of letters, numbers, and symbols that intricately trans-late Zoë's every conceivable movement and action or process.

From a concept-testing perspective, Tempest and Hyperion on Devon Island were successful. On July 18–19, Hyperion tra-versed a 6.1-kilometer circle, during which time Tompkins's col-leagues were forced to intervene only once. This was considered good! The subsequent report to NASA was optimistic. But Tompkins remembers the overall experience as excruciating. Tempest was very slow, Tompkins recalls, consuming seven hours to generate a mission plan for a twenty-four-hour traverse.

"I'll never forget: First, my colleagues are waiting half a day in a tent in the middle of nowhere for my planner to plan—a process that should actually only take a few minutes. And then finally Hyperion starts to move. It goes and goes, you're feeling good, it is all coming together until suddenly, something happens—a detour in its route, a bug in the code, whatever—something that signals Tempest to reevaluate and replan. Another seven hours!" The most frustrating moment was when Tempest figured out the plan for seven hours and signaled it was ready to begin only to discover that it had exceeded the starting time and therefore had to start all over with a new plan. Tompkins could still feel the eyes of his col-leagues burning into his back as he worked in the tent to reprogram Tempest and make it function more fluidly. "Talk about suffering," he laughed.

Tompkins is hoping that this Atacama OPS will be his tri-umphant, delayed reward—the defining moment when Tempest will prove itself in an autonomous, fast-acting experi-ence. He has rewritten Tempest from beginning to end for Zoë, and is hopeful that Tempest and Zoë will work together efficiently.

. . .

A NATIONAL COUNCIL ON Aging study published in 2004 in the journal *Science* compares the actions of the ant, which methodically and continuously plans its future, and the grasshopper, who lounges in the summer sun oblivious to the impending change of season, responding instinctively to whatever stimuli is placed in front of it. The study suggests that people, like the ant and the grasshopper, are often involved in two distinct decision-making processes: First, like the ant, we engage in constructing a long-term logical sequence of events, planning and replanning when circumstances change. This is our Tempest personality, which sees the world in a panoramic perspective, in large grids. When Tompkins said that Tempest thinks big, he meant it, not only symbolically, but also literally. Anything happening in an area smaller than thirty meters is beyond (or literally, beneath) Tempest's scope.

The second way humans and animals make decisions is considerably more emotional, less evaluative, and more reactive and spontaneous—like the grasshopper. Or also like a program written by Chris Urmson, a tall, bespectacled Canadian, who joined the Robotics Institute as a graduate student from the University of Manitoba in 1998. Urmson is the author of the obstacle avoidance algorithm, or "navigator," for Hyperion and Zoë.

Although Urmson is ten years younger than Tompkins, they both chose Carnegie Mellon for similar reasons: to make robots that move, primarily in space. And also because "you get to do things that are so cool here."

I first met Urmson in his tiny office, concealed in the labyrinth behind the steel walkway above the High Bay. Urmson's little cubbyhole was packed with electronics equipment and papered with posters of robots like Nomad in the

Antarctic or Sojourner on Mars. "Other robotics programs, you work with little guys. Here, we build the monsters," he said, pointing to his Nomad poster. "We drive them around and make them do cool things. Look," he waved an arm around the room, "I've got $25,000 of equipment on my desk; they're my toys. I can go downstairs into the High Bay and play with stuff worth millions. It's like Lego, except ten times bigger!"

Urmson first worked on Nomad in the Antarctic in 2000. "I was the person involved in getting Nomad's arm from a stowed position to deployment. "Basically, when it saw a rock that was interesting, its arm would come out. It had a camera on it, and it would take a measurement. That's what I did. But my real interest is in making robots drive over things, as well as around them. That's where I fit into the LITA program."

"Isn't that what Paul does?" I asked.

"Paul looks at the broader picture. He establishes the way-points," Urmson replied, reminding me that Tempest can only think and see in a grid of thirty-meter increments, a little less than half a football field long. For increments of less than thirty meters Urmson's navigator takes control. Some roboticists refer to the navigator as the local planner, while Tempest is the global planner.

"Look, if you wanted to go across the campus, Tempest would direct you, landmark by landmark, or waypoint by way-point, to the building you are seeking. This is also called 'bread-crumb planning.' I follow the crumbs, one by one, going around or overtop the obstacles—chairs, doors, what-ever. That's the navigator—me." Like Tempest, the navigator is constantly reevaluating in response to up-to-the-minute data supplied by the robot's eyes, or nav-cams. But the navigator can only see, at the most, a half-dozen meters at any one time, which means that there's a navigation no-man's-land between the scope of Urmson's navigator and Tempest. The ant and the grasshopper—the big and the small picture—have yet to be united.

TO UNDERSTAND EXACTLY how Zoë works, let's begin at the basic operational level. A robot, in this case Zoë, receives a command that directs it to move to a particular location, travel in a specific direction, or possibly perform a task of one sort or another. The command originates with a human, but is initiated by a navigation program or algorithm based on information supplied by a stereovision system—two video cameras, or "nav-cams," located about midheight of the robot, pointing downward from where the wheels touch the ground.

But Zoë won't see the terrain or the obstacles in its path as a human might, although it does go through an analysis that results in actions similar to those a human might take. As part of a complicated sequence, Zoë's navigator correlates or matches images, pixel by pixel, from the left and right cameras. (A similar "stitching" process occurs for the three-camera SPI.) Clouds of tiny points, called a point cloud, eventually converge into vague shapes, recognizable to humans but irrelevant to the algorithm. The shapes are projected on a checkerboard-like grid through which the navigator can recognize and assess the different paths the robot might travel. That's what Zoë sees. This is similar to Groundhog's cost map or Bernadine Diaz's occupancy grid for her pioneers.

As with Groundhog and the pioneers, Zoë's navigator thinks like an economist, assessing the cost of moving along each of these possible paths from grid to grid, considering the potential hazards along the way, and eventually choosing and following the least costly—the safest and most efficient. This is precisely what a human might do—instantly and spontaneously. The navigator is cautious, moving in nearly imperceptive curved arcs, for about a fifth of a second. It repeats the cost analysis, based on updated data, and moves again in another tiny arc for another fifth of a second.

Other robot vision systems, like the Red Team's Sandstorm, may use sonars or lasers instead of digital cameras, but the cost analysis sequence and the choice of paths and arcs are very similar, according to Chris Urmson who, lured by Whittaker, joined the Red Team to write a navigator program. Urmson's Sandstorm navigator is considerably different from Hyperion's or Zoë's in other ways, however. The navigator for Hyperion and for Zoë assumes as a baseline that it knows nothing about the world and that it is going to discover it all spontaneously through its cameras, triggered by directions issued by the science team. It stops, looks, calculates, and responds with caution.

Sandstorm, on the other hand, will know two hours in advance exactly where it will be going, as well as the basic route of how it will get there—information supplied by DARPA. Waypoints, what might be called a safe road zone, will be provided. The idea is for Sandstorm to navigate the corridor as fast as possible and not get into another vehicle's way. "But it is OK if Sandstorm gets banged up a bit," said Urmson, "as long as it makes it to the finish line—first!" Zoë, on the other hand, has to be protected at all costs. On Mars the finish line is survival.

Urmson's Zoë and Hyperion navigators are examples of the way in which sophisticated programs develop and make vital contributions—a window into Manuela Veloso's campaign that roboticists share all data. The navigation systems for Zoë and Hyperion, as well as Spirit and Opportunity and Groundhog, are structured on a platform called Morphin, developed more than a decade ago by Reid Simmons and used for Grace and many other robots, including Nomad.

To extend the hereditary connection, the navigation algorithim for MER is called Gestault, developed by Mark Maimone, a senior research scientist working for NASA at the Jet Propulsion Laboratory. Maimone's mentor at Carnegie Mellon, where, like Urmson, he earned his PhD in robotics, was Reid Simmons, and the platform he utilized to design Gestault is Morphin. "The dirty secret of software engineering," young roboticist David Thompson told me, "is not about writing software; rather it is about tooling what somebody else has written to meet your needs."

Zoë's navigator is linked to the vehicle controller, which connects the motors on the vehicle, one for each wheel, and coordinates speed and direction. (A controller is software that interfaces with hardware.) The fact that Zoë and Hyperion are on wheels, and not on two legs, like Asimo, or four paws, like Aibo, doesn't alter the basic superstructure of the way robots "think" and work: A navigator processes information from data received from cameras and translates it into directions fed to a series of controllers, which, in turn, enables wheels (or legs) to move appropriately. For Zoë alone, there are more than three-dozen software programs, most of which are not nearly as complicated as Tempest.

In addition to a vehicle controller for Zoë's four motors, there's a controller for each of nine cameras (in addition to the three SPI and two navigation cameras, there are two cameras under Zoë's body, a sun tracker camera, and a camera on the flourescent imager [FI]) and a pan-tilt unit (PTU) so that the

SPI can capture panoramas. Controllers connect to each scientific instrument, such as the fluorescent imager and the spectrometer. There are also software programs called "managers" which interact with the controllers and planners. Managers focus on a particular task. The science manager deals with science autonomy software, for example.

The executive connects all of the managers. You can think of the executive in a robotic system as kind of a vice president of operations. It doesn't make decisions regarding when and where the robot travels or what it does, but it organizes and coordinates with Zoë's many major programs and systems, including the navigator, the health monitor (Zoë's self-diagnostic tool), the vehicle controller, and the instrument manager (which coordinates mission specifications and science actions).

If all is working right, the remote science team in Pittsburgh will request a reading from the spectrometer, for example. The instrument manager will power up the spectrometer. That's one controller program or code. Then calibrate it—another program—then point it in the right direction, take the spectra, collect the data, and transport the data to a predetermined file—each requiring a separate low-level software program.

Technically, the actual writing of these programs is not difficult for a competent programmer. The challenge, during the writing and/or the debugging phases, is in imagining how things are supposed to happen while factoring in the unexpected. Powering up the spectrometer could take longer than usual, for example, or the instrument you are powering up could fail to start—inevitable glitches bound to happen sooner or later that can't be accounted for in simulation.

Unlike Manuela Veloso's little Aibo robot dogs, which are cute and irresistible, Zoë and Hyperion, essentially carts on wheels, are creatures with which we are less likely to bond. And while roboticists often scoff at the notion of emotionally connecting with their robots, even the Aibos, they are aware of the fact that the programs they write for robots are based on their own personalities and eccentricities.

When I first discussed the National Council on Aging study with Paul Tompkins, he wasn't surprised. You write the software using human models, he said, yourself included. "The more you learn about robotics, the more appreciation you have for human beings and animals—how people think and move in a coordinated way." He considers the software he is writing part of his legacy, just as a writer or philosopher hopes that his thoughts and words will, in some measure, change an aspect of the world.

Before Tempest, mission planners had not been necessary, for robots had only been required to operate in small, contained areas. But now, with Zoë, we were moving into another era when robots would begin to think for themselves, traverse great distances, and attempt to do science. The landscape of the future is evolving. Wettergreen intends to allow Zoë to log 50 autonomous kilometers using Tempest in 2004 and 200 kilometers in 2005.

ZOË, GROUNDHOG, AND SANDSTORM represent different ways in which robots think and navigate. Remember that Manuela Veloso's Aibos, Qurios, and Segways are guided by color codes rather than waypoints and bread crumbs, a delicate system, hostage to venue, vulnerable to variable lighting conditions. After the high school students at the American Open in 2003 tampered with the light switches, Veloso's student, Scott Lenser, labored through the night to recalibrate the Aibos so they could compete the following morning in RoboCup. This was not a problem which Veloso or Lenser were willing to live with.

Two years later, Scott Lenser unveiled an algorithm allowing a robot to adapt automatically to changes in lighting. In a demonstration, Veloso shows the robot first addressing and moving the ball in the bright lights of the laboratory, then in a normally lighted office, and finally in the dark hallway leading from lab to the office. This is the result of six years of research on Lenser's part, working under Veloso, and the substance of his PhD dissertation. That is only part of Lenser's work. The same algorithm was able to detect differences to various flooring materials (rubber, cement, or carpeting), while performing particular behaviors, like turning in place. Adapting to and performing on these various materials is now possible, Lenser says.

Now Veloso's robots are much smarter than before—able to detect an environmental change and adapt to it, Veloso said. Rooted in her 1992 PhD thesis and expanded by PhD student Mike Bowling in his own thesis a decade later, this new way of thinking begins with what Veloso calls a "playbook." Bowling designed a selection of various scenarios or "plays" based on his study of RoboCup action. During the match, Veloso's robots watch an opponent, evaluate its actions, perceive strategies, and subsequently choose the most appropriate countering scenario from Bowling's playbook. This is similar to how humans act; while we are not exactly programming, we have our own portfolios of responses to different situations, based on experience, which we employ as needed. Lenser took a cue from Bowling's playbook idea in designing his adaption software.

Zoë and Hyperion adapt to the environment from the ground up—a data/signal sequence in which its navigators and/or planners continuously collect information about the environment and rapidly update and respond, then update and respond again. Veloso's approach to autonomy is different in that her software allows a robot to classify the environment, overall, and perform in consort with the scenario or situation it perceives.

Veloso's robots, in other words, will recognize and evaluate the combined conditions of terrain, weather, time of day, and, possibly, other variables, and then adapt by triggering a predetermined set of actions. "Understand," said Veloso, "that my robots don't have the concept of 'hallway,' 'office,' 'lab' in their mind as humans might." They know, rather, how to recognize environments—hallway, office, lab are perceived differently—and select the appropriate action based on that environment. Tempest is responding to the environment, but in a much narrower realm—only as it relates to energy.

Veloso's Aibos could represent an important breakthrough: Rather than duplicating from the ground up the human experience, "we provide a portfolio of good things to do in any given situation—and they have to figure out, on their own, the best choice to make."

There are many caveats. For one thing, these robots are limited to the scenarios embedded in their memory banks. Referring to Gary Kasparov, she points out that the computer could, for each move, search its memory through millions of possible positions fed to it over a half dozen years by a corps of engineers. With Veloso's plan, Deep Blue might have had only 500 situations to analyze and choose from, and it might not have so easily defeated Kasparov.

Enabling robots to learn by selecting actions from a suite of plays rather than expecting them to learn from scratch provides the depth and intelligence that allows for the retention of knowledge, so that robots learn from their own experience, much like human beings. Over the past four years, Veloso's students have been attempting to write code to teach Aibos how to walk faster, for example. But recently the robots have been teaching themselves, learning through trial and error and doing a better job.

Keyed by an algorithm written by Sonia Chernova, the robots begin by walking at a pace of their own choosing for a specified time period. When the robots stop, they consult the markers on the field to measure distance traveled. They will walk again in a different manner, increasing leg speed, bending limbs in a different way, stretching out their bodies; there are fifty-four different variations from which to choose. The robots will continue to repeat the process of trial and analysis and then, like a team, share the data with the other robots in the experiment by using Veloso's team-building algorithms. "The robots figured out the optimal way for walking—and the final convergent point was 20 percent better than my students had achieved in four years. My students were never able to find that combination of values that the robots found themselves."

This is impressive in itself, but Sonia Chernova in a subsequent conversation corrected the way in which I initially described her work and put the immensity of the task in startling perspective. The idea, she said, of fifty-four variations from which to choose is accurate, but also misleading. "It sounds like there are fifty-four possible walks to choose from, whereas there are actually fifty-four parameters in each walk, each of which can have up to hundreds of values. The difference is important since if there were fifty-four walks, we could just try them all one by one and pick out the best. With fifty-four parameters to work with, there are literally billions of different walks. The goal of the learning algorithm is that it needs to test out the smallest number of different parameter combinations (walks) while trying to find the fastest walk it can. There is no way to actually test all the possible walks, there are just too many!" So in order to make real progress, one might deduce, there will be a time when humans will have to trust their robots to think and learn for themselves.

Veloso's realization that robots can learn on their own through experience has led to a related revelation: Robots can also learn through observation. Instead of attempting to define a chair or a door for a robot—a conceivably impossible task, given the variety of chairs and doors that exist—she exhibits in videos of her robots how people use chairs and doors. As soon as they see someone sitting, then the robot knows why a chair exists. Veloso has a video showing one person walking down a hall and opening a door and going inside. "Now the robot has a definition of entering and disappearing—and the place where you do it from becomes a door." Veloso is increasingly convinced that robots are at that point of sophistication so that hardwiring is unnecessary. "We will write the learning algorithm, while robots will find a metric of success on their own."

Veloso is careful to distinguish between the goals of the field robotics crews and her own work. "We are not working on tasks related to underwater or space; I am working with robots and how they exist in daily life, and we do not have a specified, time-sensitive mission. If my robots fail, nothing really happens to society or to me or to my grant or my funding agent, so I can make my goals very ambitious because it is OK not to totally fulfill them. But it is not so easily OK for my colleagues involved in mission-specific stuff. They are much more goal-dependent. They have a lot more to lose."

CHAPTER THIRTEEN

Fallback Positions

THE MINING CAMP IS ABOUT THREE MILES DOWN THE harrowing road Finch had catapulted us over the previous afternoon. Breakfast takes place at 9:00 A.M. after two crews of miners have eaten, the first at the end of the night shift and the second beginning the daylight rotation. The men who work the salt mine labor in three eight-hour shifts, digging and loading salt into trailers and hauling the coarse, white cargo to the port at Iquique where it is shipped in gigantic freighters. The reserve in this mine is enough to supply the world's demand for salt indefinitely.

Breakfast is usually quiet and congenial with a smattering of jokes and a brief discussion of the plan for the day. The food is tasty, bursting with carbs and fat: Delicious homemade bread with jam, greasy sausages, and slices of cheese are standard fare. The scene reminds me of the mess hall of my summer camp when I was a kid, for it is warm inside, compared to the chill outside, and the cacophony of the background, with music and laughter in the kitchen, provides a brief but welcome contrast to the isolation of the base camp, seemingly far from civilization.

Arriving at the mining camp, some folks carry a towel or a personal hygiene kit and duck into the bathrooms immediately before or after eating. Ample rolls of toilet tissue are stored in the trucks. The rolls of tissues are like trophies, removed from the truck and stacked at the end of the mess hall table. Periodically we will slide out of our seats, walk around the table, snatch the toilet tissue and disappear, and then, after a while, reappear, returning the roll to its position of prominence. David Wettergreen seems to judge when the crew is ready to return to base camp by watching the toilet tissue disappear and reappear on the table. When it looks like no one else will be snatching one of the rolls, Wettergreen stands up.

The team is congenial most of the time, with only a few minor squabbles and personality conflicts. Good behavior is normally the rule on field trips, but there have been glaring exceptions. In Antarctica with Nomad, "there were two guys who got into a nasty ongoing conflict over what 'autonomy' means," Mike Wagner tells me. "The first guy insisted that Nomad autonomously found meteorites based on his definition of autonomy and the other guy disagreed with the first guy's definition of autonomy. They went at it until we went home."

Although everyone is congenial now in 2004, I can sense a subtle, mounting wave of tension—a collective and pervasive nervousness gradually infiltrating the air. Zoë is designed to function as an integrated unit, with one system reliant on another system. And most everyone on the robotics team is responsible for a part of the system that makes Zoë effective. For Wagner, navigation and vision and Tompkins's Tempest, the planner. Anything pertaining to the operation of the mechanical or electrical functioning of the robot falls on Teza's and Heys's shoulders. Everyone knows that these few remaining days of shakedown to get Zoë functioning as advertised are key to the upcoming science operations the following Monday morning.

Thus, everyone is anxious to check out their particular system on Zoë and eliminate lingering problems so that the OPS runs without a hitch. Or, at the very least, without a hitch for which they are directly responsible. This is especially critical now, since Zoë had not been ready on time for adequate testing in Pittsburgh. No one is comfortable with how far his project has progressed, nor confident that the systems will perform up to par, most especially the youngest and least experienced member of the field team, Dan Villa, responsible for the position estimator.

Simply put, the position estimator (PE) tells the robot and all of its supporting systems where it is in the world. This information is essential. Neither Tempest nor the navigator can plan or map with accuracy without a starting point. And neither system can reevalute and update, as designed, without precise information concerning how far and how fast Zoë has traveled.

The PE gathers information from encoders that count how often each wheel turns; from a sun sensor that records the position and angle to the sun and from which heading is inferred; from a fiber-optic gyro, or FOG, that measures the heading of the vehicle; and from a potentiometer, or POT, that measures wheel angles. All of this information and much more—every morsel of data supplied by Zoë's myriad of sensors, which includes acceleration, distance, direction—are fed into a filter that munches it all together and provides an estimate of where the vehicle is. The PE updates and communicates position data ten times per second.

Prior to this point, no one seemed to pay much attention to the PE in the planning meetings. Hyperion's PE had functioned efficiently on Devon Island. Matt Deans, a Robotics Institute grad now working on the LITA project from his new home at NASA Ames, had updated the software after the first LITA OPS in 2003, making it ready for Zoë. The job of debugging the new version of the software and integrating it into Zoë fell to grad student Dan Villa, a tall, lanky, Seattle native. Since Villa was the least experienced member of the LITA staff, I assumed that the PE was a low-level program that required some attention, but not a great deal of concentration. I was wrong. No one could have predicted that the PE was going to change the entire scope and breadth of the OPS.

For Villa, the LITA opportunity had emerged out of the blue. He had come to Carnegie Mellon's one-year master's degree program almost as a lark. Robotics was interesting, but he was not driven or obsessed by it, as were some of his classmates. He'd be in and out quickly. But then David Wettergreen offered him a research assistantship—free tuition, a stipend, and another year of study. So why not? Villa had never been entrusted with anything as challenging and stress-producing as a NASA-sponsored project, and he was feeling rather inadequate. After joining the LITA team, he told me, "I looked around and asked myself, 'What am I doing here? I feel lost, like I am a fake!' "

Wettergreen was confident Villa could do the work, primarily because there was not a lot of work to do. The PE had not been a problem on Hyperion. As with the vision system, the PE had worked well in simulation and in the brief testing period on Flagstaff Hill. I had observed part of that testing period in the early spring when Wagner and Wettergreen were attempting to connect Hyperion with Vijay Singh at NASA. I remember meeting Villa that day because I appreciated the uncomplicated definition of the PE he provided: "The thing that tells the robot where it is." There was no reason to assume that the PE would not be in perfect operating condition. But, quite simply, it was not functioning adequately now that it was needed.

With Mike Wagner struggling to bring clarity to the vision system and with the PE completely unreliable, most of the men sitting in the mess hall that morning share a silent deep-seated trepidation: That the science operations will be a disaster—and all of their efforts individually and collectively over the past two years since the LITA project was conceived, could be for naught.

Perhaps no one is more anxious than Paul Tompkins, whose precious Tempest will be ineffective without a fully functioning PE. And since he will receive his PhD and leave Carnegie Mellon in a few months to work at NASA Ames, this may be his last opportunity to give his undivided attention to Tempest and Zoë.

Both Tompkins and Wagner understand the pressure on Villa and the embarrassment Villa is experiencing. In the Arctic in 2001, Tempest was such a disaster that it polarized the entire camp. In protest, one rebellious contingent staged a permanent work stoppage. Those choosing to continue with the Tempest experiment carried the ball. "Nobody was happy," said Tompkins. "It was a dark time."

To their credit, no one on the LITA team publicly voiced any complaint or criticism toward Villa. Everyone understood that robotics is frequently frustrating, a gauntlet of trial and error and unyielding disappointment, as Tompkins had experienced with Hyperion and Wagner was enduring now. But Villa could not help feeling ashamed. "I felt horrible, like everyone was thinking that I had ruined their experience."

But this morning, after breakfast and back at the base camp, three short days before the OPS would begin, even their always calm leader and anchoring spirit has the jitters. Wagner and Heys watch as Wettergreen struggles to find a place on Zoë for a white plastic slab—called a "white reference"—which resembles an oversized cutting board or serving platter. "It is 98 percent white, and believe it or not, it cost $1,500."

Zoë is equipped with a VNIR—a visible near-infrared spectrometer, which measures energy in order to assess minerology. The white reference determines how much sunlight or energy from the sun is there to begin with before the spectrometer readings occur. During last year's OPS with Hyperion, the spectrometer was carried in a backpackable unit, separate from the rover. Now it is attached to Zoë, but Heys has not established a place for the reference. To calibrate existing sunlight a person holds up the reference and points it at the camera.

"I'd like to dispense with it and get the $1,500 back," says Wettergreen.

"Let's just see where it will fit," says Heys. "We'll use Velcro to anchor it down."

"That's our problem," Wettergreen replies, "we don't have enough Velcro on this robot." In last-minute situations, roboticists will inevitably employ what they sometimes refer to as the "Velcro solution."

"OK," Wettergreen says, opening and closing drawers and muttering to himself, "so where's the damn Velcro?" The area around the three OPS tents is in as much disarray as is Zoë's corner of the High Bay.

"I've got my master's degree in Velcro engineering," says Wagner, overhearing the conversation. "I can find Velcro anywhere and anytime." He immediately produces the roll of Velcro, concealed under a plastic container directly behind Wettergreen.

While Wettergreen and Heys are engaged in Velcro engineering, Alan Waggoner is staring down at Zoë, as Jim Teza and Mike Wagner lift Zoë's lid in order to debug one of Zoë's computers, which suddenly doesn't want to boot up. "Zoë is making a funny noise," Waggoner says.

"I don't hear anything," says Heys, now finished with the Velcro, "but then, too, I keep imagining I see Zoë moving around in the middle of the night."

"I think that's called a dream," says Wagner.

"I am not dreaming that this weather station is dead," says Finch, pointing down at a white box with a bunch of dials, resembling a portable apartment refrigerator. "I don't know what happened, but all of the data [are] lost." The science team had requested a weather station to do temperature and climate experiments. "I guess we can do without it. We have no choice," he adds.

"Just another thing going wrong," says Heys quietly. "We are getting down to the wire, and I am getting more and more worried."

Jim Teza has gone into the supply tent to find a monitor to hook up to the computer to try to find out why it won't boot up, and Wettergreen, who has followed, says he wants him to go into town with Finch to pick up some supplies.

"But I want to be doing some of my own work." Teza has developed new lithium batteries that will double the power on Zoë. They cost $25,000, a significant investment, and he had been hoping to test them on Zoë today. "This is not a good time for that," says Wettergreen.

"There never seems to be a good time," Teza answers. "I can't get my work done."

"It can't be helped," Wettergreen replies quietly.

Teza says no more. Teza is dark and tall, a ponytailed, hippie type about Tompkins's age, who hardly ever speaks. And when he does, he is so soft-spoken his colleagues are frequently asking him to repeat what he says. Teza has a 1996 degree in electrical engineering from the University of Pittsburgh, and a passion for literature (mostly, but not exclusively, science fiction) rather than robotics. "But I was always interested in computers, only I wanted them to do something in the physical realm instead of staying so stationary and static. It was a disappointment when I realized the limitations of robots. I was naïve." Now, he says, robots are catching up to his intellectual expectations—slowly. Teza works at Carnegie Mellon on a project-by-project basis, as do most engineers in this software-driven science hub—not particularly conducive to stability. Teza tried programming to enhance his capabilities and found it "tedious and exhausting." Zoë is Teza's fourth robot and this is his second foray into the Atacama.

He produces the Samsung flat-screen monitor to connect to one of Zoë's four computers, along with a keyboard to use to debug Zoë. The computer sounds like it wants to boot up, but the monitor is dark. Zoë's six computers were purchased at the start of the project early in 2003, while the monitor was purchased right before they packed up for the Atacama—a light, flat screen selected for clarity and portability. For a while, there's hope:

"It will be all right," says Wagner.

"Give it a little time," says Teza. "The computer is slow."

Now we were all standing around Zoë and waiting. "This is not going to happen," says Wagner, now looking over the manual that comes with the monitor.

"What's wrong with Zoë's computer?" asks Waggoner.

"I think there's nothing wrong with the computer," says Wagner, "that a little tinkering won't fix. But it is a few years old, and the monitor is too sophisticated to display the data. Too new for the job. We didn't consider that when we bought it."

"Thank God we are in Chile and not New York," says Waggoner. "Here we can find equipment that is very low-tech."

"Now I am really worried," says Heys.

"Oh boy," jokes Waggoner, "Zoë is having a nervous breakdown."

"That makes two of us," Heys says.

LATER THAT DAY, Wettergreen and I are driving to the airport to pick up Vijay Singh, the NASA programmer who wrote the executive. I ask how he thinks things are going. "We had a clear set of objectives to start with, but much less of an idea of how everything was going to work once we got here. By experience I know that things will go wrong. The real test is how good we are at improvising and dealing with our difficulties and how to recover from them. We are always talking about our 'fallback' position." Wettergreen is referring to how the roboticists adjust to changing priorities and the inevitable decreasing expectations precipitated by technological uncertainty.

"How do you think everyone is holding up under the pressure?"

"These are all good people, and I can count on them to make individually good decisions. But what I need them to do is to make collectively good decisions. Everyone has to do their work and, at the same time, pitch in and help others. We are all interconnected," he says.

"Are you nervous about what is going to happen on Monday when the OPS begin?"

"I feel responsible. NASA has poured a lot of money into this project and expects us to get results, and I don't want to disappoint them."

He suddenly downshifts into second gear, stomps the throttle to the floor, and whips the Toyota around a turn. I fly up in the air, bounce against the door, hold my breath, close my eyes, and give thanks for seat belts and roll bars. So even the cool, calm, and collected David Wettergreen unleashes a wicked, crazed frustration when behind the wheel of a truck. Wettergreen's aggressive driving makes Finch seem timid. Here again the roboticist's syndrome—explosive wildness as a balancing point and escape valve from the stationary intensity of the keyboard and display. Or, in this case, an anxious and ecstatic moment of relief from the stress and fear that all of his plans may go awry at the critical moment.

Wettergreen doesn't slow down when we pass the mine. In fact, he cranks the throttle. "This is not going to turn out to be a perfect field trip, I know that." The throttle is nearly to the floor, the desert whizzing by in a burnt red blur. "That is what you learn early in robotics. Some projects are not going to work perfectly, and you are definitely going to have to adapt and work around the systems that are not functioning. These field experiments are lessons in prioritizing, making choices. You learn to make the most with what can be accomplished."

We are nearing the airport, and Wettergreen is gradually slowing down, thank God. Failure in itself is only a setback, but not necessarily irreversible, he observes. "A negative experiment can sometimes be a good thing, as long as you are aware of your mistakes and figure out how to make them more positive the next time."

"That seems to be the lesson of robotics," I say. "It's the 'never give up' scenario."

Robotics, he answers, can't be compared to disciplines like engineering or surgery. "I mean, there's no plan or manual. Nothing is structured or expected. Most of the things you try don't work right away. It requires constant diligence to make a system work, and sometimes, even after all of the time and effort, blood, sweat, and tears, they don't work at all. That's the nature of the beast."

AT THE AIRPORT, I decide to go into town with Finch and Jim Teza. Wettergreen asks me to copy a shopping list he has drawn up in his notebook—at least I had thought it to be a notebook until I got a better look at it. Black and spiral-bound, it is actually an artist's watercolor journal. It is made of fine, high-quality, unlined pages. I flip through it quickly. There are lists and notes everywhere in very neat script or clear block letters. Wettergreen often inserts a square or cube in front of each point to bullet the items on his lists. "I like to draw lots of pictures," he says, "especially when I am in meetings or sitting around and thinking." I turn the pages as he is speaking. Once, paging through his notebook, I spot-

ted a series of delicate watercolor landscapes. German farm-
land, painted on a recent trip. "My mother was an art teacher,"
he says. "My brother is a potter."

Many of the computer geeks here maintain diaries or
hand-written logs alongside their computer records and code
writing. It makes me feel good—and still relevant—that pen
and paper are respected tools of the trade and are not going to
be squeezed into cyberspace, at least for the moment. And
they all have books they are reading—novels mostly—except
for Teza who is more eclectic. On the way into Iquique, Teza
discusses John McPhee and Diane Ackerman, authors I
admire. He peppers me with questions about the lives of writers as we wend our way along the winding highway that follows the shoreline into town.

Teza is probably the most distant and reticent member of
the group—a soft-spoken loner who rarely complains. Despite
their disagreement earlier in the day, Wettergreen is sensitive
to the individual personalities of each member of the group—
perhaps Teza especially. A few days later, during dinner at the
mining camp mess hall, Teza was missing. It had been a long
day. Everyone was tired and hungry; it was nearly 9:30 P.M.,
but Wettergreen got up from the table without eating, jumped
back into the truck, crashing through the rutted road in the
pitch black to find out what had happened to Teza. They
returned together a half-hour later. Teza had fallen asleep in
his tent and wasn't aware when everyone mustered in the dark
for dinner. "Teza is so quiet," someone remarked, "you don't
know when he isn't there."

The security guards at the supermarket in Iquique are
wearing bullet-proof vests. Finch calls this "the intimidation
factor," the Latin American macho experience. Finch's mother

is an English teacher and his father is a physician. This is his second year working with the LITA team. After last year's OPS, Wettergreen arranged for Finch and a companion to work at the Robotics Institute for a month. Finch was graduating college this year and hoping to study robotics at the RI, beginning fall 2005.

I am aware of the country's history of violence and political unrest, but firearms and protective vests seem so unusual in relation to what I have observed in Chile so far. The people I meet seem gentle. Most of the people in service positions— from supermarket employees to airline reservationists—wear neatly pressed uniforms, with a quiet and beaming essence of pride. The stores are much cleaner and better organized than in parts of the United States, and even the poverty-stricken sections have a dignity and neatness to them.

In the supermarket, we heap two dozen five-gallon jugs of water and 200 rolls of toilet paper into three shopping carts, but no one seems to bat an eye at checkout. Stocking up for the desert is evidently a normal procedure; there are many scientific expeditions in the Atacama. We find a duty-free store in a modern indoor shopping mall and purchase an inexpensive monitor for $80. Next, we locate a store similar to Home Depot, where we buy bolts, cargo straps, and lumber. As we hop from place to place, Jim Teza wonders aloud why we are getting all of this material, especially the lumber, and why we have been sent on an errand without being told the objective. We find out later that the materials are for Stuart Heys in response to Wettergreen's request that he build a roof rack on one of the trucks to transport Zoë to the OPS site. We have also rented an SUV we call "the soccer-mom van" to use as our base camp and wireless network hub in the field.

The fact that Finch and Teza are unaware of the reasons for the materials and that they are somewhat annoyed by their ignorance is indicative of the difficulty of maintaining a stable group dynamic here. Working together on the same project for a long period of time and now living together, a certain intimacy develops in which everyone assumes that he is totally aware of what everyone else is thinking and doing. But that is often not the case. When I ask Alan Waggoner and his assistant Dave Pane if Zoë's inability to see or know where she is disturbs them or their work, they both look up at me in surprise. "That's for the roboticists to deal with," Pane says. "We've got our own stuff to worry about."

CHAPTER FOURTEEN

The Desert Makes Us Wacky

SEPTEMBER IS MIDWINTER IN THE ATACAMA, SO IT IS quite chilly in the morning. Sometimes I wear a sweatshirt, a down vest, and a windbreaker on top of a turtleneck. We gradually peel our layers as the sun burns off the frigid air and dries the heavy morning dew. In a place where rain is unheard of, the chilly morning wetness is disconcerting, but tame compared to the sudden, bone-shaking cold, as low as twenty degrees Fahrenheit, in the ink-black darkness of the night.

By 7:30 A.M., two days after I arrive, Mike Wagner is tending to Zoë's vision system and helping Dan Villa understand and correct the position estimator problem. Wagner doesn't mind switching his focus. "You can't fully test the navigation system until you get position estimation working." Meanwhile, Stuart Heys is working away on Zoë's roof rack, what we are now calling Zoë's "chariot," to transport Zoë to the site of the Monday morning OPS. Using a hammer and handsaw, Heys works slowly, a piece at a time, while chatting with Jim Teza, who is wiping down Zoë's solar panels.

Of the three large tents in the camp, the first and smallest is used primarily for supplies. Here's where Jeff Moersch has dumped his equipment before heading out on an expedition to the south. Moersch, a geologist from the University of Tennessee, will be using two spectrometers. One called a visible near-infrared spectrometer (VNIR), identical to Zoë's but equipped with a device that allows his instrument to be laid on a rock to eliminate sunlight as a variable. Having an identical spectrometer is essential for comparison in ground truth checks and balances. Moersch has a second spectrometer, a TIR (thermal infrared spectrometer), that works on the same principal as the VNIR—measuring the amount of energy being reflected in different wavelengths from the rocks—but is much more sensitive to lower amounts of energy being produced. Both spectrometers will seek signs of chlorophyll and isolate and measure the mineral composition of rocks and soil. Now that Moersch's tent is vacant, Alan Waggoner closets himself inside, making notes and studying specimens and mumbling to himself. Preliminary tests of his fluorescent imager are not going well.

The other two tents are larger, suitable for sleeping eight, although no one sleeps here, unless dozing out of sheer exhaustion in front of a laptop display. Each tent contains a table on which we stack computers, monitors, and power strips for laptops. Chairs are at a premium. Some of us sit on boxes, our laptops crammed tightly against our crotches. The power generator is located between these two tents, as is the satellite dish connecting us with the remote science lab and the rest of the outside world.

Paul Tompkins, Vijay Singh, and Dom Jonak, with whom Tompkins, Waggoner, and I had traveled, have immediately burrowed into one of the two tents, sitting side by side with their laptops, as if rooted there. With Zoë incapacitated, they can devote time to updating and debugging code, integrating programs, and running simulations. Tompkins is pretty much a one-man Tempest show, while Singh is obsessed with the Zoë executive.

Working with both programmers "is like watching a tennis match," says Dom Jonak, who is assisting both. Singh and Tompkins are running their programs repeatedly, testing integration and waiting for something to go wrong—"to break," as Jonak puts it. "It is either Tempest or the executive, and whichever one breaks first, they fix it [by rewriting the code] and then run the programs again and again until something else breaks." They will go back and forth like this for days in simulation. Later, when they finally got into the field, they knew there would be plenty of other "breaking" points before the two programs would be totally integrated.

Singh is content at this moment being inside the tent rather than outside in the desert. Jonak, the second youngest member of the LITA team here, has few expectations about what a field trip is supposed to be, so he is not disturbed about remaining in the base camp. But he is somewhat perplexed with the state of the robot in general, at this point, so close to the all-important OPS. "I assumed that the low-level software would be working when I arrived at camp, since the people who were responsible had plenty of experience doing this sort of thing and they were on the scene. There would be some routine debugging— perhaps a day or two—and then we'd be ready for the science OPS. That's what I figured, but this doesn't seem to be the case."

Tompkins is patient, but itchy. Periodically he will look up from his laptop and comment, as if the sound of his words reassures him that everything will come together. "The planner can't do anything until the state is straightened out." The PE is often referred to as "state" or "state estimator." At one point, Dan Villa announces that the PE is fixed. Tompkins is excited. He jumps up and down and shakes his fist in triumph. But soon Villa announces a false alarm.

"Back to the drawing board," Villa says.

Later, Wettergreen enters the tent, "How is your stuff going?"

"Sucky—badly." Villa says he is in a box and can't fight his way out.

Neither Tompkins nor Singh nor Jonak want to be sandwiched behind the walls of the tent for too much longer. But at this point, they have no choice but to run the executive and Tempest through simulations and wait for failures to occur.

"We look for the little things. The executive has bugs. Tempest has bugs. The goal manager has bugs. This goes on and on," says Tompkins, "until it makes us crazy."

Tompkins and Singh have been trying to get together and do this integration work for months. It is ironic that they have traveled 6,000 miles just to be confined inside a tent in the middle of the desert, shivering in the early morning, sweating in the afternoon, and shaking late into the night to finally coordinate their efforts.

This is yet another example of the difference between doing a "concept" project, which is what Carnegie Mellon does, rather than the real thing: The approach and preparation can be less rigorous because they are not as clearly defined. And as Manuela Veloso noted, they don't count as much as the

MER project. Expectations are always lower—which, however, doesn't necessarily make the task easier or less worrisome to the people involved.

"I hate this," says Tompkins.

"But," says Singh, "we are a lot better off today than yesterday."

"What day is it, actually?" Dom Jonak asks. "It's like I have been here forever."

"Saturday," says Dan Villa. "I know that because I have been counting down to Monday."

"I am worried about Monday," Paul Tompkins says. "I can't stop thinking about it."

AT NIGHT, TENT LIFE is exceedingly intimate. We layer on clothes, squeeze into confined spaces, and periodically chat while hunching over laptops. The windows of the tent are zipped shut. We can't move around outside without flashlights. Waiting for dinner at 9:00 P.M., the group is hungry and giddy, speaking in English, Italian, Spanish, and whatever other language we think we know, intermittently. For some reason, the "F" word, absent during the day, abounds at night with time on our hands. As the time nears for dinner, Mike Wagner begins to hum a patriotic march, something from a parade. Tompkins and Heys pick up the tune, humming louder and louder. They break into song. Then everyone in the camp in both tents is singing and laughing and stomping feet in a simulated march. At the end of the march, they cheer—and then go back to work. "The desert makes us a little wacky," Alan Waggoner says.

Waggoner will wax philosophically from time to time,

especially in these close quarters. He is comfortable airing his own insecurities when the opportunity arises. In a conversation with Dom Jonak, Waggoner once confessed that he didn't really know—is not sure—what he wants to do with the rest of his life. Waggoner owns five acres of land near Santa Fe, New Mexico, for which he has paid dearly, but now, all of his friends and former colleagues at the Los Alamos Proving Grounds, where he once studied, have retired or moved away. So he's not quite certain what to do with the land. And he envies his son who, in his mid-twenties, is still trying to find himself. Last year he was a Grand Canyon tour guide and this summer he's clubhouse manager for the Eugene, Oregon, minor league baseball team. Waggoner wants his son to get a real job, but by the same token wishes he himself had more "fun" in his life. "I'd like to do something more adventurous." He stops and squints at Dom Jonak, through wire-rimmed glasses. "Dom, you are looking at me like I am crazy."

Jonak had earned his degree in 2003, serving as a teaching assistant in five courses. In his senior year, he briefly considered graduate study, but a friend tipped him off to the LITA project's need for a software-savvy person. David Wettergreen immediately recognized his potential and hired him on the spot.

"Dom," Waggoner repeats, "what are you thinking?"

"Well, at this point you should know what you are doing with your life," Jonak tells him. "You are rather old to start to plan the future now."

Jonak had never heard of Carnegie Mellon University. Growing up near Boston, he had always intended to study where all the important work in computer science was being done, supposedly, MIT. But his father convinced him to at least take a look at this Pittsburgh place with a reputation for "in the trenches" education. In 1998, Jonak went to a "sleeping bag weekend" for high school students in which you meet computer science freshmen and sleep in their dorm room in your sleeping bag, like "a little puppy at the foot of their beds," he laughed at the memory. He liked the campus, which is only a short walk from the larger and more sprawling University of Pittsburgh on one side, and nestled at the edge of the lush public Schenley Park with running trails, swimming pool, and golf course, on the other. You can see a green corner of the park from the wall of windows in the atrium. The Carnegie Mellon folks were laid-back, and Dom appreciated the always-present connection between computers and real life.

As the LITA project progressed, Waggoner would become periodically annoyed with Jonak for his perceived arrogance— not atypical of the bright young men who are trusted in responsible positions to do vital work, but at that moment Jonak's observation tickled Waggoner. "You are absolutely right, Dom," he says. "You would think I would have a direction and be aware of the goals in my life and how to fulfill them by this time. But I am not so certain I am that together. I prefer being out in the desert and away from the office, that's one thing I know."

Neither Jonak, nor any of the other bright young men of the Robotics Institute, mean to seem arrogant; they would have been embarrassed and apologetic if they had known they were being perceived that way. Jonak's personal home page provides a very sweet glimpse of a young man just now coming out into the world. In a portrait with classmates in his first robotics course, Jonak, slender and baby-faced, with a jutting chin, a longish nose, and a thick shock of wavy brown hair, is wearing his gray Carnegie Mellon sweatshirt. He is beaming at the camera with boyish satisfaction—a twenty-year-old in 2001, just beginning to recognize his possibilities and harness his capabilities in a productive professional fashion.

The home page captures him with his family—his parents; his Golden Lab, Princess; various birthday parties; a pumpkin on Halloween. There are images of his first car, friends and parties, cap-and-gown high school and college graduations, an especially endearing one with his elderly "babcia," grandmother in Polish. Jonak's family emigrated to the United States from Poland via England. He is a British citizen, although he grew up in the Boston area. Jonak does not now know whether robotics will be his life's work, but to this point robotics has opened heretofore unimaginable opportunities, he says.

"WACKY" MAY NOT BE a scientific term, but it adequately captures the atmosphere and Alan Waggoner's prevailing mood. Midnight before the beginning of the OPS, Alan Waggoner and Dave Pane, his colleague from MBIC, are looking at rock samples that had been sprayed with fluorescent dye.

"I can see lichen," Waggoner observes, examining the rock in the dark with a flashlight. Lichen is the most visible and accessible sign of life present on the surface of rocks. In some areas of the Atacama, rocks with lichen are everywhere, while in other areas, lichen are completely absent. Waggoner recognizes salt and gypsum in these samples, and he wonders how to figure out what the energy generated by the fluorescence actually means. He is concerned because the lichen in his rock samples aren't radiating as much energy as expected. "It might be because the rocks have been stomped on by all of us and then run over by Zoë for the past ten days," he speculates. "Or it might be dead lichen. It might be that the pigment in these rocks is so old that there's nothing left to radiate."

He lifts a rock up to his eye and studies it carefully, scrunching his gray brows under his tanned, wrinkled forehead, like a jeweler surveying a diamond. Lichen coloration varies widely, from bright yellow and orange to more common greens and grays. "These rocks are pretty." He studies some more, then suddenly bursts out in song: *"I feel pretty, oh so pretty."* This to the tune of the song "I Feel Pretty" from the musical *West Side Story*. Waggoner goes on like this for a while as he studies the rock and ponders the problem, until he becomes engrossed.

Developing the fluorescent imager (FI) had been a real challenge because Waggoner had assumed at the baseline beginning that a measurable fluorescent signal required total darkness. Because of this, he and Wettergreen had had long discussions about the feasibility of conducting the fluorescent tests at night. Wettergreen had been resistant, although agreeable, explaining that Zoë at night could not traverse great distances; it would have to rely on battery power; sunlight would not be available for recharging. Zoë traversing at night would also be dangerous, both to the robot and the field team. "We could do it, but I would like to find another answer," Wettergreen had said. They had experimented with a shroud or curtain, but that was not really a satisfying solution, since you could never guarantee that light would not leak through somewhere. The dilemma caused part of the delay Stuart Heys had been complaining about.

Then one day Waggoner overheard a colleague discussing a new high-intensity xenon flash system he had read about and suddenly he had "a brilliant idea. I said, 'Holy cow! Suppose we use intense flashes of light while collecting fluorescent images." The intense flash—1,000 watts of full-spectrum lighting—would create a mass of sudden fluorescence. The dye measurement is made during a ten-microsecond period, which is a hundred-thousandth of a second—compared to one full second when fluorescence is recorded in the dark, Waggoner explained. "We tested the flash system in the laboratory and then out in the field and the results were incredible. Park and Elmer, the manufacturer, never imagined their flash used in that way."

The FI was then designed and put together by his staff in a few months and installed on Zoë by Stuart Heys only hours before the shipping deadline in August. Time for testing was minimal, but the MBIC staff was confident the FI would work. Now, however, the fluorescence is giving off a weak signal, not strong enough for accurate measurement. And Alan Waggoner cannot understand why it is happening. Nor does he know what to do about it.

As Waggoner ponders the lichen in the tent, jotting notes down on a yellow legal pad, Dave Pane, quiet, round-faced, and heavyset, is squinting into the display of his laptop, typing with ferocity. His fingers fly over the keys for many minutes. Pane stops and ponders and then types again. Singh, Tompkins, and Jonak move in and out of the tent, conversing from time to time, but Pane's concentration is unwavering.

Of all the LITA team members on site, Pane is perhaps the most—and least—prepared for OPS. A graduate of the University of Pittsburgh, he has never been out of the country or camping until now. Pane carefully studied the preexpedition materials distributed by Wettergreen and made certain that he was more than ready for all variations of weather and terrain.

He arrived with a duffel bag stuffed with underwear and outerwear for winter and summer, plus his own tent, "because I snore and I didn't want to disturb anyone."

But nylon walls do not contain the vibrant honks of exhausted men sleeping in the cold solitude of the desert night. As a light sleeper, I can attest to the fact that Pane, by far, was not the only member of the field team with snoring problems. There were points of time in the middle of the night when the desert floor vibrated—a chorus led by Pane.

Dave Pane working in the tent in the evening in the Atacama. *Courtesy of Alan Waggoner.*

DAVE PANE IS THE most nervous and insecure of anyone on site. At forty-four, he is a family man with three teenage children to support and a job as programmer that depends on research projects that Waggoner and others solicit. Pane had been part of MBIC for thirteen years, but he frequently moves from project to project. Unlike Waggoner or Wettergreen, there is no tenure in his job. If something goes wrong here, Pane has more to lose than Heys or Villa or Tompkins, all of whom are young and not yet family providers.

The FI is actually a black-and-white camera with a fixed focal length lens; the focus on the target is achieved by adjusting the height of the FI up and down until it finds the "sweet spot"—a sharp image of the precise target on the ground to be studied. The FI works by taking a series of images to pick up fluorescence. Two filter wheels aid in this: the first positions a filter in front of the high-intensity pulse flashlamp to determine the color or wavelength range with which the scene will be illuminated. This filter is often called an excitation filter or, as Dave Pane puts it, the "illuminating" filter. The second, in front of the camera lens, determines the exact wavelength range the camera is going to see for the data it will collect; these are often called emission filters. Since the dyes' fluorescence characteristics are known ahead of time, the excitation filter puts the illumination in the wavelength range for maximum absorption of a certain dye. The emission filter lets the camera see only the wavelength range that is expected to be emitted by that dye. Months ago, the job of writing the software—"controller"—to interface all of these complicated components fell to Dave Pane.

Pane completed programming for the imager far in advance, but he had taken a gamble he now partially regretted by going on a family vacation to Florida for two weeks before Zoë had been shipped. The Pane family vacation had been scheduled long before he had been asked to accompany Waggoner, and he had worked overtime to make certain that his part of the job functioned smoothly before he left. He had assumed that, during his absence, his colleague, biologist and instrument specialist Shmuel Weinstein, who was building the FI, would go through all of the scientific protocols determining the imager's actions and positions. Pane had also assumed that since he had completed his part of the software interface that the roboticists were also on track. The integrating software from the robotics team was crucial to making his controller function successfully. The controller connects with the instrument manager.

Theoretically, the instrument manager on Zoë was to send a message requesting a specific scientific experiment to Pane's controller, which would in turn direct the imager to follow the specified instructions. Each instrument was linked by a controller to the manager, which sent directions and authorized each individual order or task. When Pane left town, the instrument manager had not yet been completed, nor had any of the controllers, except for the one for which he was responsible. But it was, he had been assured, in the works and being written by Trey Smith. It would all come together in the Atacama, if not before. At least, that is what everyone assumed.

So Pane wrote an awkward and elementary program that would mimic the directions of the instrument manager in order to make certain his controller software functioned adequately. Flying off to Disney World, Pane expected that his counterparts at the Robotics Institute would fulfill their part of the bargain. Smith was very busy and was unable to complete his program, but he planned on working with Pane to fix any bugs discovered in the instrument manager when he arrived in the Atacama, just as Vijay Singh and Paul Tompkins were working together now.

Smith felt the pressure, but he appreciated the fact that his boss, David Wettergreen, was not as obsessed or set in his ways as other faculty. "Nobody on this project is negative, and no one will invest time and effort tearing you down. I don't know exactly why we all work together so well, except for the fact that Dave Wettergreen has tremendous moral authority—and it isn't because he has power. Everyone respects his opinion, and his leadership style is subtle and hands-off and calm, so when there's a problem he can step in, say a word, and clear the air."

This lack of negativity—Wettergreen's ongoing and unrelenting optimism—is obviously genuine and very appealing to young people. Engineer Chris Williams, twenty-four, says he works hard for Wettergreen—harder than he has ever worked for anyone before—because "he's such a nice guy."

Once in the desert Williams lost a gas cap from one of the Toyota trucks. He admitted that he had an annoying habit of misplacing things, and as a kid his parents would give him a hard time about it. "I went to Dave all sheepish and said, 'I screwed up and lost the gas cap.' 'Oh, OK,' he said, 'we'll get another one from Hertz.' It was no big deal. David is focused on the goal of the project—and that is where he wants you to be; you don't want to let your teammates down."

Which is perhaps why Trey Smith did not balk or complain when Wettergreen broke the news that Smith would not be going to Chile. Smith had been scheduled for thirty days in the Atacama, but at the last minute David Wettergreen told Smith that he must stay in Pittsburgh and finalize the software programs by which the data sent by Zoë were accepted and processed by the science team. So suddenly, the instrument manager, the project on which he had been working for so long, was in limbo.

"Some of us were aware of this development—and some weren't," Smith said.

DAVE PANE HAD returned to Pittsburgh from Disney World for only one night before jetting off to Chile—a bizarre transition, as if going from one fantasyland to another, except the latter was real. Prepared to work the moment Waggoner, his boss, arrived in camp, Pane immediately and carefully tested his controller software from start to finish. It worked. The autofocus went up and down and the filter wheels rotated on command. Pane was ready to synch up his software with the instrument manager Trey Smith was working on. He was surprised and disappointed when he discovered that most of the instrument controllers and the instrument manager were unavailable in time for the OPS.

"My whole focus in the weeks before the desert was to get the imager to accept images and transfer data—and it did that." Pane kept his own counsel, but had never imagined the fractured way in which the OPS would begin until he found himself knee-deep in the confusion and the disarray. "I thought this group of robotics guys were in control, that they knew what was going on. I thought that we were going to get down here in the desert and the rover was going to do all of these incredible things. The reality was a rude awakening."

As it was for Stuart Heys. When Heys signed on for fifty days in the desert, he didn't realize what a personal toll it would take. "In the desert, you lose your sense of the world. You forget what day of the week it is after a while, and no matter how big the desert is, you are never alone. When the sun goes down, you are crammed in a tent to try to keep warm. And it gets lonely. As an engineer, I don't have any code to write at night to keep me busy, so I surf the Internet and write e-mails to my girl-friend. She's miserable being alone, constantly complaining. Mike is trying to get me to pull the plug. But you can't break up with someone over the e-mail. 'Just give me your computer and I will get rid of her for you,' Mike says."

Advance preparation was lacking from the MBIC side as well. In retrospect, Waggoner admits that perhaps he should have reviewed the protocols Weinstein wrote in Pittsburgh and asked Weinstein to revise and simplify them before the OPS, for when he examined them in the desert for the first time he realized they were cumbersome.

Waggoner had immediately decided to streamline each protocol and, over the few days before the OPS, he had made considerable headway. But during the first few days of the OPS, Pane and Waggoner must constantly adjust and recon-figure to get them right. Polishing protocols is beneficial, but the process causes frequent delays.

During this tedious trial-and-error period, Waggoner real-izes that the results from protocol to protocol are inconsistent. He receives data from one protocol that makes sense given the procedure and then data from another protocol on the same procedure that is off-the-wall. Pane and Waggoner work on and off trying to understand the reasons behind the inconsis-tencies with frustrating ups and downs—frustrations similar to Wagner's struggle with the vision system and Villa's with the PE. At one point, Waggoner is convinced that the filters are faulty. Light is spilling over from one shot to another. But minutes later, he concludes that the filters are fine but the filter wheels are in the wrong position. "Are the filters moving?" he asks Dave Pane.

"They are moving on command," says Pane, studying his computer display. He can see the imager functioning.

"I can hear them moving," says Mike Wagner.

"But you can't hear and see at the same time," says Waggoner.

"I can hear if you can see."

Waggoner crawls down under Zoë and squints up at the imager. "I can be your eyes, if you can be my ears."

Pane goes through a series of commands, but as far as Waggoner and Wagner can see and hear, the imager is functioning.

"What do you think the problem is?" says Mike Wagner.

"Is it my program that's bad?" says Pane.

"Or my program?" says Wagner, who has written a small part of the interfacing software.

Later in the day, Pane accidentally signals the filter wheels to go to the same position twice instead of following the protocol sequence. Once it goes to the right position and once to a different position—from the same signal. "Wait," he says. "I think I have found something." The wheels are inconsistent because of a bug in the code. "I wrote a couple lines of code, and we had the whole problem straightened out," says Pane.

Accidents and seemingly unrelated events sometimes become irreplaceable keys to scientific advancement, as Alan Waggoner would soon demonstrate.

CHAPTER FIFTEEN

Peeing on a Rock

At the designated site the afternoon before OPS, Zoë is lowered from its chariot, powered up, and tested. The wireless network is established on the roof of the van—the base station—to communicate with Zoë at a discreet 100-yard distance. The pan-tilt unit directing the SPI is tested. Everyone is ready to start. Excitement and trepidation are in the air. But no one really knows what is going to happen. How can the OPS be OPS if Zoë is not functioning? The irony is palpable. The LITA team is not ready, but that dismaying fact doesn't seem to matter. The OPS is here and they will try to go through with it.

Late at night, the remote science team sends the plan for the following morning. Most everyone, except for Wetter-green, is huddled in the tents by then, sleeping—or trying to stop worrying and get to sleep. Wettergreen downloads and prints up the science plan, and it is ready for us to review when we wake up. We have breakfast in the emerging dawn and hit the road back to the site of the OPS where Zoë has been sleeping.

You can't see anything as you creep through the darkness. The headlights barely pierce the blackness. The ruts and rocks jarring the trucks make dozing impossible. The eeriness, the emptiness, the silence is all pervading—frightening and exhilarating and isolating and extraterrestrial. Your thoughts are crazy and in this blackness your identities become scrambled. Is Zoë the robot or are you the robot and is Zoë you? Are we all robots here in the Atacama—or are the real robots back in Pittsburgh fooling us into believing that they are the scientists? Or is Pittsburgh actually Mars? This is what Mike Wagner postulates to Stuart Heys who replies, "Wake me when it's over. I think the whole damn OPS are a bad dream."

We bounce our way in the Toyotas, thirty minutes of rocking and rolling and knee banging to the site. The remote science team has sent the first set of instructions directing Zoë on a particular set of maneuvers with coordinates. Zoë is powered up in the emerging dawn. After a while it is traversing slowly across the desert floor.

Inside the soccer-mom van, there are four guys in fleece, funky hats, and filthy clothes with dirty, unshaven faces. Wagner, Pane, Wettergreen, and Finch are sitting in perfect stillness with notebook computers on their laps, huddled together in a tangle of wires crisscrossing over and around their heads like spaghetti, all laptops connecting to the network they had previously established. Peering intensely into their laptop displays, they doggedly tap away on their keyboards.

Wettergreen and Finch come and go from the soccer mom van, but Wagner and Pane are the two key figures in this mix, with Wagner controlling and directing Zoë by sending keyboard commands through his laptop, and Pane fulfilling the function of the instrument manager and transmitting signals to Zoë to initiate the fluorescent imager. Pane can also monitor the functions of the FI and intervene when necessary.

The atmosphere in the van is intense. All you hear is the tap-tap-tap of keyboards and periodic crackly conversations on the walkie-talkie between Wagner and Pane in the van and Waggoner, Teza, and Heys with Zoë. For Pane and Wagner, typing and programming is very awkward and uncomfortable—and more than a little un-Mars-like in the soccer-mom van. But the situation signifies the unwritten irony of this entire experience: How to carry forth the façade of Mars relevancy and the Atacama OPS generally when not a lot of what you are counting on and expecting to work works?

Wagner is especially uncomfortable and unhappy. To access his keyboard and type, he has been forced to jam his laptop against the steering wheel in an awkward angled position with his wrists pressed into his chest and a forearm and elbow, to which he must continually apply sunscreen, out the window. "You look uncomfortable," someone says.

"I am actually sick," he replies, sniffing and blowing his nose. "Who is going to believe I got a cold in the middle of the desert?"

In the van, there are long periods of silence as the four men work intensely. Finch is trying to get the spectrometer to work; it won't. Wettergreen is e-mailing Cabrol and other science team members. Periodically Pane will wave a woolly mammoth feather duster and lightly brush the red dust from his computer display and keyboard. Periodically Wagner will blow his nose or sneeze. Periodically Wettergreen will hike up the hill from the van to peer at Zoë, now over the horizon. Periodically Zoë will creep out of range of the wireless network and Wagner will fire up the van and follow along until the base station picks up the

Zoë signal. The wind whips the red dust up into swirls knee-high from the ground. Fantas and Fracs, the snacks of the day, supplied by the cook at the mine, are cracked open and devoured. Fog wafting from the ocean hovers in the sky.

Somehow—and no one mentions this seemingly astounding fact—the LITA roboticists seem to have created a miracle recovery. We are proceeding with the science experiments without a blink of the eye, as if nothing untoward had occurred and the hardware and software tissue of Zoë were functioning and meshing flawlessly—or at least it might seem so to the outsider. "Zoë is perking along," Stuart Heys remarks. The rest of us are silent.

UP ON A HILL, one afternoon, perhaps a half-mile away from the soccer-mom van, Alan Waggoner is lying in a dry gulley, bathing in the sun. Waggoner's arm is extended outward toward Zoë. Waggoner's fingertips are touching its wheels. Man and machine are connected. Little did Waggoner realize when he first boarded the plane to Chile that he was going to spend most of this time on the ground side by side or under Zoë, an unrelenting, sometimes intimate, and often annoying liaison.

Waggoner loves the desert and is happy to be cut off from the administrative responsibilities of the MBIC, but he is also frustrated. While the FI protocols have been slowly and steadily streamlined, he continues to be puzzled by the reasons why the emanating fluorescent signal is so weak. His dyes are usually powerful triggers of energy and heat.

Over the past few days, Waggoner and Pane had narrowed down the Shmuel Weinstein protocols to the bare bones. They sprayed the dyes and tested how long it took for the dyes to penetrate the samples—from one to fifteen minutes—and then they took a series of images with the FI. They eventually concluded that waiting ten minutes after spraying dye, allowing time for the dye to penetrate, yielded the best results so far. But the results are still not good enough for Waggoner. His dyes are not radiating sufficient energy. He can't understand why, no matter how hard he concentrates and experiments and postulates, using Dave Pane as a sounding board, until another accidental observation provides the Eureka moment for which he is hoping.

Time passes. The day goes by, sometimes droning and sometimes galloping. It is now early evening. Waggoner struggles to his feet. He needs to think in silence without the chatter of his colleagues and the crackling of the walkie-talkie—and also without the looming presence of Zoë, haunting him. He also needs to relieve himself. Waggoner has devoted most of the day to polishing protocols, collecting rocks for ground truth purposes, and pacing around, muttering, trying to solve the riddle of the lichen.

He walks for five minutes, almost in a daze. The desert does not provide a lot of places for privacy, but he doesn't mind. The exercise feels good after lying on the cold rocky ground for so long. Eventually, he slips behind a low dune surrounded by a few rocks. He looks out over the vast expanse of rusty-red emptiness dotted with patches of dazzling white salt flats, shadowed now in the impending dusk, all the while going over in his mind the protocols he developed earlier in the day. He continues to ponder. The litany in his head goes on and on.

But then, minutes later, when he returns to Zoë, he is smiling—and he immediately adds an extra step to the protocol, a sequence that had not been anticipated. Now, he tells

Pane, the terrain being examined would be sprayed with water before the dyes are dispensed by the FI. And they discover almost immediately that by spraying water on the specimen and waiting a few minutes prior to applying the dyes, the image that they capture generates more energy—more than enough fluorescence to measure.

How did Alan Waggoner discover the importance of spraying water on a rock before imaging it? "I noticed how bright the rocks got," Waggoner says, "after I peed on them." The water activated the surface so that the dye could penetrate more easily and deeply. Later, Waggoner added vinegar to the water spray and he experimented with other liquids, including alcohol. But the water and vinegar spray, because of the acid it contained, worked best.

NOW, WHEREVER ALAN Waggoner goes, he carries around a white bucket with his water spritzer and his notes and rock samples to be shipped back to Pittsburgh for ground truth purposes. Periodically after spritzing, Waggoner will close his eyes and lie back, totally relaxed for ten minutes while the water sinks into the area to be tested and the FI is activated. "I'll just sit in the sun and wait," he says. "Maybe this is what it is like to be dead."

Sometimes Teza and Heys are lying beside him in the sun, waiting for Zoë to move or go through its FI protocols. Sometimes so many of us are lying beside Alan Waggoner that if anyone spotted us from a distant hilltop, they would think it a mass suicide. Teza and Heys are with us to make certain that Zoë doesn't kill itself. It is not as if Zoë is out of control, although it is certainly not in control, either. Mike Wagner in the van, scrunched by the steering wheel, his fingers awkwardly clawing his keyboard, is the heart and soul of Zoë at the moment. Wagner keeps Zoë in control—to a point.

The waiting for Zoë to move from place to place, to take panoramas and other images and spectra, seems never-ending. It is like the hurry-up-and-wait syndrome in the military. Instantaneous rush and whirl followed by prolonged periods of limbo.

Even after Pane's software adjustment, which fixed the filter wheel, the imager frequently refuses to flash the necessary fifty times (the protocol they have established), and when that happens, they have to start from the beginning. "This software stuff is a real demon," says Waggoner.

He is not, to say the least, a computer-savvy scientist. "Someday this robot and imager might all work wonders, change the world, but not now, not for a long time. I wish we had done this for a week on Flagstaff Hill and then had two more weeks to analyze the data and make adjustments before coming down here. I think everyone wishes that about many aspects of Zoë, too, though. Clearly, there's been too much on-the-job adjusting." A miscalculation, he admits. "But we rallied," he says.

Dave Pane is both hero and casualty of the campaign. "The first few days of science OPS, we were changing everything by the hour—or minute," Dave Pane recalls. "During the day, when things weren't working, I continued to modify my program every chance I got, following Alan's suggestions. And at night, I would go back to the tent and modify my program even further. I knew we needed to move forward and the instrument manager was not in any shape to do it, so everything and everyone was relying on me. I had not anticipated how important my program would be—and all the pressure I had to work under was terrible—all of this disorganization."

He was shaken up and wary of what might happen to him in the future. "As I sat there at night, I just wanted to work as hard and as long as I could because I thought my job was on the line. If we don't get some good data, I thought, Alan is going to either kick me out or make me stay in the desert forever. Either way, I felt like I was a big failure." None of this was true, from Waggoner's point of view. Waggoner understood the challenge Pane faced and appreciated his efforts. Pane was a standout contributor, but even into the OPS a month later, Pane's nerves were at breaking point.

UNLIKE DAVE PANE, his partner in the van, Mike Wagner was not feeling guilty. Wagner was simply attempting to adjust to current reality and fulfill LITA responsibilities to the unknowing remote science team as efficiently as possible. This was not a crisis situation from his point of view; it was quite simply part and parcel of the frustration of working with robots. Wagner was adjusting and downsizing just as he had learned to do repeatedly since joining the Robotics Institute. He and Wettergreen, the two leaders, had to accept the fact that autonomy, the grand objective, was no longer a priority—or even a distant possibility at the moment. Pure functionality—literal survival of the field campaign—was at stake.

Wagner knew he had to find a way of servicing the remote science team and facilitating results one way or another. And he succeeded in that regard simply and rather ingeniously. Wagner was maneuvering Zoë from his laptop from place to place by feeding Zoë's navigator a series of images or maps he had saved from Devon Island when Hyperion and Tempest were being tested. The images he utilized were flat so that the navigator was led to believe there were no obstacles in front of it, and it could go anywhere it wanted in a straight line—or actually where Wagner signaled and directed it, as dictated by the science team. So Zoë was traveling blind—no autonomous navigation whatsoever—just Mike Wagner at his laptop as its remote operator. Stuart Heys, Chris Williams, and Jim Teza became

Pamela Hinds, Stuart Heys, Jim Teza, and Zoë waiting for a software fix.
Courtesy of Alan Waggoner.

Zoë's shadow; following along and making certain Zoë didn't destroy itself by falling over a cliff or crashing into a rock.

THE OPS IS NOT only a test of technology, but also an excruciating evaluation of the patience of every member of the LITA team on-site, Heys, Tompkins, and Singh especially. Waiting is the operative word.

Once, Waggoner, Wettergreen, Heys, Teza, and I, all of whom are following Zoë, are asked by Wagner, via walkie-talkie, to hide from Zoë's cameras so that Zoë can take a panorama without recording our images. We climb a hill and disappear behind a row of boulders. At first, we sit quietly soaking in the sun, alone with our own thoughts, feeling the wind whipping against our cheeks, until Alan Waggoner says, "Lee, I have a very sharp knife in my hand. And I can't cut the lichen off this rock because the lichen is so hard."

Thinking that this has some scientific significance, I ask: "Is this lichen different than the lichen you encountered here last year?"

"Last year I never tried cutting it with a knife or a rock."

"Is there a reason to try it now?"

"I have nothing better to do," Waggoner says.

There is silence for perhaps five minutes, as we contemplate.

"Mike," says Wettergreen into his walkie-talkie. "Just checking on the progress of the experiment."

"We're debugging before the panorama."

"Are you debugging the imager or the SPI?"

"The SPI. I'll let you know how much time is left when we are finished the debugging."

Along with "waiting," "debugging" is the operative OPS word.

Ten minutes go by.

"I am starting to understand why lizards sun themselves," says Wettergreen. "It's cold out here in this wind, especially when you stand up. But when you lie down on the ground in the sun, it's pretty comfortable."

"Lying here, waiting for the debugging," says Waggoner, "I

suddenly imagined that I was on the beach in the sand. I even have my buckets, like I have on the beach. And spritzers to cool me off. It's like being at the seashore."

"Except it hasn't rained here in a 100 years," says Teza.

"But the ocean is with us. Taste the fog," says Waggoner. "Walk a couple of miles and we'll see it from 4,000 feet up."

"OK," Mike Wagner interrupts our conversation over the walkie-talkie. "We should be under way with the panorama."

"I always wonder about the 'should' part," Alan Waggoner comments. "I'd rather he'd say, 'we are now under way with the panorama.' "

Fifteen minutes pass. "Now we're really under way with the pan," Mike Wagner says.

"That's what I thought," says Waggoner.

"What's the status on the fluorescence?" asks Wettergreen into the walkie-talkie.

There's another long pause. No answer. Waiting and waiting. "Radio check," prompts an anxious Wettergreen.

"Let me talk with Alan," says Dave Pane on the walkie-talkie, finally.

"This is Alan."

"I want to go over the filter protocols one more time. I reprogrammed everything." He reads off the steps. "Correct?"

"Now it's correct," said Waggoner.

"I have a little more debugging to do," says Pane.

"The panorama is taking an unusually long time," says Mike Wagner.

"What's going on?" says Wettergreen.

"I am not sure what the issue is. It should be done by now—long before now."

"What's your plan?"

"My plan is to continue to look into the problem with the pan until Dave Pane is finished debugging the fluorescence. When Dave is ready with the imager, I propose we abort the pan."

Waggoner and Wettergreen have a long talk about the rocks on the top of the hill. The conversation goes on for perhaps twenty minutes, then abruptly, Waggoner asks: "What kind of things were you interested in as a kid? Robots? Mechanical stuff?"

Taken by surprise, Wettergreen thinks for a long time. "My mother is an art teacher," he starts slowly. "I did a lot of art projects. Worked in a photography studio for a while. I worked two years as a veterinary technician, then worked at the Rochester Zoo cleaning up after the elephants."

"So, veterinary medicine?"

"Yes, I considered it."

"I don't know why this is taking so long," says Mike Wagner on the walkie-talkie, "but I plan to keep pursuing it."

"Can Dave Pane start with the fluorescence? Is he ready?" says Wettergreen.

"He thinks he needs another five minutes."

"Could be all of this lichen is dead," says Alan Waggoner. "That would be a shame."

"How would we know if this is dead or living?" asks Heys.

"We need to call in a lichenologist."

After a minute, Heys understands that Waggoner is joking and he laughs.

Wettergreen is becoming very restless. He takes his baseball cap off and on, stands up, and peers down the hill at Zoë, then sits down again.

"Can I walk down and take the shell off the rover while the fluorescence is going on?" asks Heys.

"If we could only put a water spritzer on the rover, it would help us a lot," says Waggoner.

Pause.

"Want to put a water spritzer on the rover?" Wettergreen asks Heys. "For next year?"

"I swore I would never put hydraulics on the rover," says Heys. "Something filled with evil liquid. But I guess water is OK."

Silence.

"For some unknown reason our wireless network just went off," Mike Wagner says over the walkie-talkie. "We are changing locations in the soccer-mom van so that we can reconnect."

"Well then, I guess," says Wettergreen, "we can now abort the procedure."

They stand up, slapping their pant legs and initiating clouds of red dust, yawn, and stretch.

THE MISSION PROCEEDS in the same fractured way with prolonged delays in the field as Pane continues to modify the imager program in place of Zoë's instrument manager, and Waggoner further fine-tunes the protocol. Mike Wagner continues to fool Zoë into believing it is on Devon Island traveling in a straight line with no obstacles in its path. When the OPS ends four painful days later, Tompkins returns to Pittsburgh and Singh to Mountain View, California, without the opportunity to employ their code on Zoë.

CHAPTER SIXTEEN

Downtime

MIKE WAGNER RETURNED HOME BRIEFLY TO ATTEND a friend's wedding. A week later, he was back in the desert, motivated and energized, but worried that the bleak situation he left behind would prevail for the second OPS. It was hard to believe—and it was depressing—that the low-level basics on Zoë were not functioning. He was surprised at the massive fallback positions that they had to take to make the OPS work, like driving Zoë with Devon Island images—pretending it was autonomous. Wagner was not yet panicking. "It's like being on a plane, and the plane is late, and there's something you have to do where you are going. But you don't get angry or make yourself crazy because there's nothing to do about it until you get there."

He sympathized with Dan Villa, saying that Dan was only partly to blame as nobody had had time to work with him, and with Singh and Tompkins, who flew halfway across the world just to be squeezed together for ten days in a tent. But over the next ten days, between the first and second OPS, the roboticists gradually began to catch up with their proposed agenda. In four concerted days of trial-and-error testing, Wagner discovered that the nav-cam focus was distorted at the outer edges of the lens, not in the center where he had been concentrating his attentions. He finally fixed and recalibrated it.

It took another four days to discover that Hyperion's PE, now on Zoë, had not functioned nearly as effectively as they had assumed. Strangely, it was receiving accurate information from Zoë's sensors but not believing or accepting it, which compounded the problems with Zoë. During the 2003 OPS, Hyperion had pinpointed its position based almost wholly on the commands it was receiving from its other systems, such as the navigator. Thus, Hyperion was able to intuit where it was without the PE. Realizing this, Wettergreen and Villa fashioned a makeshift software instrument so that localization was possible.

Gradually, with many of the low-level nuisances out of the way, the second OPS began coming together. Near the end of the final week in the field—nearly sixty days after Wettergreen arrived to scout with Heys, Teza, Wagner, Villa, and Finch—the procedures between Zoë and the remote science team were finally working (at least sporadically) in a manner that was closer to their original conception. The remote science team could provide Zoë with coordinates and Zoë would respond, traversing the desert, capturing the landscape with its SPI cameras and operating the fluorescent imager, just as they had hoped would happen.

By the end of the OPS, Wettergreen could claim that Zoë had traversed fifty kilometers, autonomously. Better yet, the remote science team could ask Zoë to return to a place it had previously visited and, give or take fifty meters, it could do it. But the success of these long traverses assumed that the terrain was flat and that there were no complicating obstacles along the way, because even though Zoë could now see, the navigator was not ready for prime time in the Atacama.

Dry streambeds and railroad tracks on elevated foundations tortured and tantalized Zoë. Originally written for the flat Antarctic, the navigator had, says Dom Jonak, "no sense of pitch." Venturing into a streambed, for example, Zoë's nose pointed downward. But the navigator, looking ahead, interpreted the opposite upward slope as a wall—a barrier. Perplexed, Zoë would back up and try again, see the wall again, and retreat, maneuvering back and forth repeatedly, stalled with indecision.

Sometimes Zoë would read the streambed differently, as an obstacle to go around, and follow it endlessly, creeping farther away from its ultimate objective on the other side. Mechanically, Zoë possessed both the power and the mobility to go up, down, and over ravines, streambeds, and railroad tracks, but intellectually, Zoë was stymied.

When Zoë was stuck like this, the roboticists disabled the navigator and joy-sticked Zoë a few meters forward over the obstacle and then reenabled the navigator: A simple recovery for the navigator, but one that now became an annoying mystery to the executive, which, said Jonak, took great offense. "'Why is the robot moving?' the executive wants to know, 'I didn't tell it to move—and I give the orders here.'" The executive behaved like an overwhelmed human being: It refused to continue under the present circumstances; in other words, it stalled, went into a state of limbo, stuck in an infinite loop, and refused to recover until it was restarted. Jonak explained how the executive had been expected to work:

A module called the "health monitor" was to receive signals from all the modules in the system, and if it detected something wrong, it would send a message to the executive, notifying it that a fault condition existed. "Something wrong" could mean an obstacle blocking Zoë's path with no maneuvering room to bypass it, as in the case of the streambed or railroad tracks.

In an ideal situation, the executive would direct the navigator to deal with the problem. The health monitor would report back to the executive that the problem had been solved. The executive would then instruct Tempest or another planner to generate a new plan, refocusing on the original objective. The planner would send the plan back to the executive and the executive would notify the navigator and Zoë would proceed forward. This would all occur in an instant, under the best of circumstances. But if Zoë was still in the streambed, the navigator could become confused or cause the executive to become confused, necessitating a new plan from Tempest and leading to a new start-up and annoying delays.

The NASA programmers had written the executive so that it would think quickly and logically and, like humans, be able to make a number of safe assumptions about the world. Vijay Singh and his colleagues had designed particular response actions, or "models"—ways for the executive to respond to as many scenarios as they could imagine that would interfere with Zoë's performance. But they invariably missed a few, leading to confusion and, from time to time, aborted missions.

Trey Smith, who eventually made it down to the Atacama, wrote an emergency program that would do the work of the executive. They called it the Rover Executive Classic because it relied on proven code—nothing new or fancy. The name was also in response to the ongoing debate between the programmers who loved Apple products and those who opted for the Windows operating system. The new OSX operating system for MAC had recently been introduced, but you could still run OS9 programs under a "Classic" category. The name and the spirit behind it stuck. Jonak's description of the "Classic" sounded like he was watching a racecar. "It is lightning fast. Just tell it where you want it to go and it gets you there in a straight line, no time wasted." This made Jonak smile.

"Our rover is insanely complex," said Jonak, and some of the modules, like the executive, are vastly more complicated than they need be for the job. But in this atmosphere of spontaneity and artistry—not to mention anarchy—Singh had had his own personal objective.

Singh was attempting to develop a general, all-purpose system executive for use in spacecraft and other advanced rovers being developed at NASA Ames, and he was attempting to test certain applications by interfacing it with Zoë. This is how scientists cobble together support, like jigsaw puzzles, from various projects. So in many ways, it may have been too complicated for Zoë—at least in the rover's present iteration. Scaling up Zoë for the future, the executive would be well suited.

Tempest, too, may be considered overkill for Zoë, but Paul Tompkins required support and data for his dissertation. This is one of the many reasons Zoë's budget is only $4 million while the less-sophisticated MER is nearly $1 billion. Researchers at Carnegie Mellon and other universities have multiple projects and responsibilities, including teaching, and many students to mentor—researchers who are also devoted to their own projects. Further complicating the use of Tempest was the need to register the satellite data on which Tempest relied—"register" meaning locating or connecting an exact geographical marker to the DEM. Planning is useless, as in the case of the PE, without first localizing—knowing where you are.

From the very beginning, Wagner, Smith, Jonak, and perhaps others would have preferred eliminating Tempest from the OPS entirely. Zoë would function more efficiently without Tempest, they insisted. Besides, as Zoë began operating more smoothly, traversing longer distances, the field team asked the scientists to plot Zoë's routes using the high-resolution DEM data they had available. Waypoints were provided, and all Zoë had to do was drive in a series of straight lines. So Tempest was actually unnecessary. In 2005, Trey Smith produced a "goal manager" program that performed essentially the same job as Tempest for Zoë, but, similar to Smith's Classic executive, in a much less complicated manner. As for the long-range goal of operating autonomously, said Wagner, "If you rely on scientists to send us a plan, then, technically speaking, we are not going to be autonomous, anyway."

So in the end, Zoë was said to have logged many autonomous kilometers, depending, of course, on what is meant by "autonomous." The OPS was considered, technically, a success. But no one on the LITA team was particularly satisfied, for they had all made major compromises, falling back primarily on the technology of already-proven programs. The cutting-edge advances they all had hoped for had not materialized, with the exception of Alan Waggoner's device, which had sent back vivid signs of life in the Atacama that demonstrated the potential of his remarkable fluorescent imager. "Zoë's instruments were able to detect lichens and bacterial colonies," Waggoner explained in a presentation at the 36th

Lunar and Planetary Science Conference in Houston, March 14, 2005, "We saw very clear signals from chlorophyll, DNA and protein. . . . Taken together, these pieces of evidence are strong indicators of life."

At the conclusion of the OPS, Dan Villa wandered Chile and Peru for a few extra weeks, but most of the LITA crew returned home to unpack the crates and distribute the soil and rock samples gathered for ground truth purposes for analysis. Essentially, they took it easy through the holidays—November and December.

Stuart Heys felt the weight of the weariness of the OPS perhaps more than any other member of the field team. Not only did he break up with his girlfriend, but also in a way he temporarily parted company with Zoë. When Zoë arrived back on campus, Heys resisted unpacking. "Right now I don't want to look at Zoë because I realize that there's so much stuff on her that isn't perfect. And she is all dirty and banged up now from the OPS. I was hoping we would leave it in a crate for the rest of the year. There were so many things I could have done better; I am ashamed of some of the work I have done." He seemed to lose his sense of the mission itself. "This is a weird project," he told me. "I don't know what the decisions are based on, and the goals are not clearly defined— except to make the scientists happy."

Heys, the engineer, may not have been keyed in to what Trey Smith called "the two versions of reality" in many of the robotics projects developed at Carnegie Mellon. I too had wondered how decisions were made, for I would attend meeting after meeting and see how the goals and the agendas shifted week to week and sometimes daily. Conversations went on endlessly—until backs were up against the wall and something had to be done. At one point, the roboticists were committed to far-field navigation—getting the robot to see in the gap where the navigator ends and Tempest begins. How and when it would be done was continuously debated and deliberated, and then, suddenly, the concept was off the project radar, as if it had never existed.

"There's a formal agenda—what we said we were going to do initially—and then the underground agenda, which relates to what each software developer on the project is actually spending their time on," Smith observed. The decisions are formalized after they become reality. Far-field navigation was on the agenda as a priority until no one worked on it—and then it was no longer discussed and no longer a reality.

Until Smith had actually explained the process, I had not realized how unscientific the decision making was. There was a method, though—a process of spontaneous exclusion or inclusion. "It's like the black market," Smith observed. If you are part of the inner circle, you somehow know what is happening. Outside this small group, most are clueless.

The dual reality of decision making might be compared to the similar reality of bonding with the robot, I discovered. As a software engineer, Dom Jonak understood the science and technology behind Zoë's creation, yet he somehow felt, at the end of his time in the desert, that he had, in some inexplicable way, bonded with the creature he had helped to build. As time went on, Jonak said, he could intuit Zoë's limits, when it would work and why it would default, even before it happened. He was not alone. Chris Williams could watch Zoë and accurately predict its movements.

The bonding between Zoë and the roboticists was gradual and subtle. When I first became involved with the roboticists, the robots in the High Bay were referred to either by name or with the impersonal pronoun "it." During the first OPS, Hyperion was "it." And even as Zoë was being built, it was "it." But at some point in the desert, the roboticists experienced a transition. Hyperion, Nomad, and the other machines in the High Bay were not their babies, so it was a simple matter to remain distant, but for Zoë, Jonak noticed, there was a change in gender.

"Suddenly everyone was referring to Zoë in the female form—*she*. It was blatantly obvious we scientists and engineers were personifying—I had never imagined that would happen, but it did." He believes that the "hardware guys" started the anthropomorphism. "They built the robot." But then he added: "I guess I also built it—I made it think, in a way, so it is also a part of me. Zoë is all about autonomy, and when you do that, you are asking yourself, 'What would I do in this situation?' "

Despite the delight of bonding, Jonak discovered, as did Heys, that being in the desert for forty days was increasingly stressful, not nearly as much fun as he expected it to be. "We frequently fantasized, jokingly, 'Wouldn't it be terrific if Zoë just dropped off a cliff and we could all go home?' "

Dom Jonak and Stuart Heys had no intention of quitting LITA or Carnegie Mellon, but they were exhausted, for one thing, and, more to the point, generally unaccustomed to the all-consuming stress and isolation they experienced in the Atacama. There were pressure-packed experiences in college, but mostly of limited duration, such as final exams or term papers to write. And it was inside the academy—for grades—not for NASA. They had acted grown up, basically, but now their tender ages and lack of life experience were revealed when the work was done and they realized what they had been through. They seemed mildly shell-shocked. I saw the same wide-eyed numbness in Scott Baker and Zachary Omohondro after the Mathies Mine traverse with Groundhog.

Even Mike Wagner admitted that by the time the field campaign was over, he had become extremely antisocial. "I want to be by myself now, because you are never alone on a field campaign, living in close quarters and sharing a tent." Wagner's wife, Angela, told me that her husband rediscovered Pittsburgh and his friends every time he returned home from a field campaign, as if he had been away for years. The hardship had been much more rigorous on Wettergreen than on anyone else, though. "Dave was in the Atacama three different times over the eight-week duration—longer than anyone; Dave took it on the chin. But I guess that's what a leader has to do," Wagner said.

Older and more philosophical, a veteran of many field campaigns, David Wettergreen was able to view the 2004 OPS with a balanced clarity and an optimistic spin. "Remember that the objective this year was to test functional integration." He did not expect all components to be fully operational, but the basic interactions between the components were working. At the very least, they had discovered why they weren't working so that they could remedy them for the 2005 OPS.

He ticked off some of Zoë's many accomplishments. By the end of the second 2004 OPS, Zoë's computer was controlling the imager without Dave Pane's intervention, and the data gathered by the imager were transferred through the telemetry stream of the robot back to the Remote Science Lab, not to mention the fifty autonomous kilometers of traverse. With the problems related to the nav-cams and the PE resolved, Zoë had navigated between science sites, although a "seamless experience" in which Zoë followed a science plan for the day—autonomously—had not been achieved. Hopefully, that task would happen the next time.

Wettergreen knew that the navigator would have to be tweaked to remain mobile when in troughs or other similarly confusing terrain, and he was disappointed with the executive, which had not adapted well to Zoë. "It was like putting a heart into a thing that never had a heart before. You can't assume it is just going to go." Wettergreen had asked Paul Tompkins to return to Chile after the OPS. They tested the individual elements of Tempest then, successfully, although not integrated into the rover.

Wettergreen admitted that Tempest, like many of the other components tested on Zoë in the Atacama, was not ready for prime time this year and that these OPS were not the perfect testing ground for Tempest in the first place. "But sometimes the people who work with me forget that our mission is to develop and test concepts." Red Whittaker was running a race in Nevada with Sandstorm and attempting to map the Mathies Mine with Groundhog. These were specific mission targets—much easier to understand and to measure as to success and failure. Zoë's objectives are less tangible, but perhaps more critical. Zoë represents the future of robotics, decades away, while Groundhog and Sandstorm symbolize what can be achieved technologically today.

Tempest takes the concept of "planning" to another level by asking essential questions related to resources: "Is Zoë warmed up enough to function properly? Is there enough dye in reserve to conduct a full fluorescent sequence? Is there enough space to store the data that Zoë collects?" These and other considerations are part of the checklist sequence that scientists or staff members will go through before going off on an expedition, and if robots are going to become scientists, then they will require a method and a means to achieve such tasks. At this point, Tempest remains in the Mars repository data bank as part of the Mars technology program, which is funding approximately seventy different technological initiatives of one sort or another, from landing systems to communications protocols, all to be evaluated at some point for flight and developed over a period of time.

Tempest might not be the final answer in relation to Mars; it illustrates the limitations of the current state of robotics and the breadth and expanse of its potential. It is a doorway to the next level of autonomy, although not Mars-appropriate at the moment. "The things we are attempting in the desert today are not applicable to Mars for a long while—if ever." A mission to Mars, Wettergreen explained, actually looks backward, not forward. Technology is thoroughly tested—something so well known and detailed that it will not easily fail. "On Mars, Zoë would have never lasted a day—an hour! The MER robots utilized thoroughly tested technology three or four years old.

"I actually think we struck a good balance between ambition and practicality during the OPS. We pushed the system—we pushed hard—and while it is true that some of the components did not work until the last few days, if at all, it helped us understand next year's challenges. As I said, we struck a good balance. We don't want to push it so far that nothing works, and we don't want to be so cautious and conservative that everything works. We want things to break. That's when we know we have pushed the system to its limits."

This "push until it breaks" philosophy, or "failures metric," is the message of robotics everyone at Carnegie Mellon endorses, and it is very much a key component of the frustration barrier inherent in the entire Robotics Institute. Dom Jonak described it to me as their "stress-testing system": "Let it run. Keep it running. Run, run, run—until sooner or later something will break, and when it does, you stop and fix it and let it run again, until it breaks, and on and on, until it no longer breaks, and then you know you have fixed the problem—until, of course, it breaks again."

I traveled home from Chile with Wettergreen, chatting and dozing sporadically during the total twelve-hour flights between Iquique and Pittsburgh. At the baggage pick up, I witnessed his reunion with his family.

"What should we do today?" his wife Dana had asked.

"I need to go to the office," Wettergreen replied.

"You've been away this long and you have to go back to work today?"

"I'm sorry," said David Wettergreen. "Just for a couple of hours."

CHAPTER SEVENTEEN

Two Versions of Reality

THE E-MAIL FROM DAVID WETTERGREEN TO HIS ZOË project colleagues and staff caught me and many others off-guard. It began by referring to the spate of positive publicity the project had received over the previous few weeks, even though it was months after the OPS. The e-mail arrived at a time when I was involved observing other robotics projects. I had not been in touch with Wettergreen for a while.

As I read the beginning of Wettergreen's e-mail, I imagined how satisfied Wettergreen and Waggoner were at this moment. In this day and age, science and technology are anchored in positive publicity. But then, suddenly, Wettergreen's e-mail turned strange.

UNFORTUNATELY, THIS WILL BE MY LAST MESSAGE TO ALL OF YOU. I REGRET TO SAY THAT I AM LEAVING CARNEGIE MELLON UNIVERSITY.

DURING MY RECENT TRIP TO THE CMU QATAR CAMPUS IN DOHA, I WAS APPROACHED BY A MEMBER OF THE AL AKSA

MARTYR'S BRIGADE. HE EXPLAINED TO ME THE GREAT EVIL THAT THE UNITED STATES IS SPREADING IN THAT PART OF THE WORLD. AS HE CONTINUED, I WAS HORRIFIED TO REALIZE THAT MY WORK WITH NASA AND THE CMU QATAR CAMPUS IS AIDING AMERICAN OIL IMPERIALISM. HOW COULD I CONTINUE WITH THESE TERRIBLE TRANSGRESSIONS ON MY CONSCIENCE?

SO, I REGRET TO SAY THAT I WILL BE RETURNING TO THE MIDDLE EAST AREA TOMORROW, PERHAPS NEVER TO RETURN. MY BROTHERS AND I WILL CARRY THE JIHAD INTO THE DESERT SANDS. PERHAPS, GOD BE PRAISED, I WILL BE ABLE TO ADAPT THE ZOË PLATFORM TO AUTONOMOUSLY CARRY MUJAHADEEN WARRIORS AND ROUT THE INFIDEL. REMEMBER ME IN YOUR PRAYERS. ALLAH KEEP YOU ALL SAFE.

—DAVID

P.S. HERE IS MY NEW CONTACT INFORMATION

MUHAMMAD AL WETTERGREEN
SECRET AL AKSA TRAINING CAMP
MAIL STOP #37
JERUSALEM, PALESTINE 65902

At first, obviously, I doubted the legitimacy of Wettergreen's e-mail. He was, after all, the project director and its chief cheerleader. I immediately considered e-mailing Wettergreen or some of the other project members to see if the message was legitimate, but before I acted, a second e-mail appeared on my computer screen, sent directly to Wettergreen, with all members of the LITA list-serve copied. This e-mail referenced

Wettergreen's original e-mail, and communicated this caution-
ary and frightening warning—*IN BOLD!*

FROM: NASA INSPECTOR GENERAL <IG-FBI@ARC.NASA.GOV>

THIS MESSAGE HAS BEEN FLAGGED BY AN AUTOMATED CON-
TENT ASSESSMENT SERVICE AS A POTENTIAL THREAT TO THE
NATIONAL SECURITY OF THE UNITED STATES OF AMERICA.
ACCORDING TO USA PATRIOT ACT, PROCEDURE 0401, SECTION
1US3R, WE MUST NOTIFY YOU THAT THE DEPARTMENT OF
HOMELAND SECURITY WILL BE INVESTIGATING.

PLEASE DO NOT RESPOND TO THIS AUTOMATED MESSAGE -
YOU WILL BE CONTACTED SOON.

At this point, I began to believe that the e-mail was plausi-
ble. I knew that Wettergreen had recently visited Qatar where
Carnegie Mellon had established a campus, and he was planning to travel extensively in the area with colleagues in search
of sites for developing and testing other robot projects after
Zoë was completed.

And the truth is, much of the work in robotics performed
by Carnegie Mellon and most other universities in the United
States is supported by the military, especially the Department
of Defense. The technologies developed at Carnegie Mellon
for NASA could easily be adapted for fighting and surveillance
apparatus. This is a fact of life at the school that is rarely discussed but eminently apparent.

While Wettergreen seemed like a rock on the outside, perhaps he was finally buckling under the pressure to make it all
work. On the surface he had seemed completely unaffected by

the problems with Zoë, confident and articulate in stating the objective of his research. But underneath, who could really know how people will respond to an overwhelming buildup of pressure? Even the strongest and most settled personality can sometimes break and do things that are totally uncharacteristic and self-destructive. I flashed back to the scene in the airport with David Wettergreen breaking the news to his wife that, after being away for so long, he intended to go directly into the office before he could spend time with his family.

I had met Dana Wettergreen for coffee at a Starbucks close to their house one afternoon months after our brief and somewhat embarrassing airport encounter. A math and business major in college, she is a stay-at-home mom, cyber-schooling their two children. "My husband is an amazing person," she told me. "He is fair, honest, decent, and operates with integrity. He is even-keeled, patient, and not emotionally volatile."

When I told her that his robotics crew and the science team held him in great regard ("They admire Red, but they want to emulate David") she was surprised that people would recognize something valuable in being calm, patient, and methodical—not causing problems—not being that "all over the place person" that Red Whittaker is. "The Red thing is so much sexier." She confirmed her husband's claim that he never worries about the projects he is working on. The one "worry" on his mind now "is how to support all the people who are counting on him to keep their jobs."

It is interesting to see how many researchers working at Carnegie Mellon are forced to modify their orientations once they become successful. As they gain support for their research, they are necessarily elevated to a management role, losing connection to the vision that initially guided and inspired them. Their managerial responsibilities can become so all-consuming that they are forced to delegate the scientific tasks to others, so that they can keep the project and the staff afloat. Alan Waggoner and Reid Simmons admitted that they are somewhat disconnected from the day-to-day scientific substance of their projects. Perhaps Wettergreen had lost his center and was feeling overwhelmed by his responsibilities. Everyone has a breaking point, I thought.

The day after receiving the Wettergreen e-mail, I e-mailed Alan Waggoner, the senior statesman of the group, and asked if he had any insight into what was going on with Wettergreen, if the e-mail was legitimate, and he answered, in part: "Lee, I am out of it, just like you."

Now I was even more perplexed, but I e-mailed back one more time, explaining that I was out of the country at the moment at a conference; otherwise I would have walked over to the Robotics Institute and knocked on Wettergreen's office door.

A few hours later, I received another message from Waggoner. "Lee, think about it. What day is today?"

I looked at my calendar. It was April 1.

I was flabbergasted and embarrassed that I could be so out of touch; I was not in Slovenia, after all, just Vancouver, Canada. But I had not paid attention to the date and I had forgotten all about last year's April fool's gag. When Wettergreen went to the High Bay, April 1, 2004, Hyperion was missing. He searched all over and couldn't find it—until he went into the tiny office shared by Jonak, Smith, Teza, Heys, and a few other staffers. Hyperion was jammed into the room. An unidentified group of cohorts—no names mentioned—had disassembled Hyperion and reassembled it in the office, waiting for Wettergreen to happen upon the robot in the morning.

But it also occurred to me then that both April fool's jokes were symbolic of certain aspects of the Zoë project—and of so many things that seem to be happening in such a broiling technological hotbed as Carnegie Mellon, where geeks and code monkeys sustain the drive, energy, and overall spirit of the place. Both tricks were brilliantly conceived and executed (Wettergreen revealed that he had received follow-up messages allegedly from the FBI, as part of the April fool's boondoggle), as one might expect from a CMU geek.

At the same time, the tricks were inappropriate and troublesome, as are many youthful pranks. It is often difficult to remember how young and, sporadically, immature the students are. Hyperion had been damaged during its magical emergence in the tiny office, and this year many of Wettergreen's colleagues on the LITA e-mail distribution list had either been taken in like me or disappointed with the poor taste inherent in the messages. It was Carnegie Mellon in a nutshell: Youthful exuberance leading to ingenious resourcefulness and an occasional inadvertent faux pas.

Here too was the sum and substance, nuts and bolts, of their erstwhile leader, David Wettergreen—thorough and responsible, unflappable and forgiving to the bitter end. When I asked Wettergreen whether he knew the identities of the perpetrators and, if so, would there be consequences, he told me that he could guess the people responsible for the prank, but that he had no interest in confirming it or following up with recrimination. "I would have preferred that it had not happened," he said, "but I am not opposed to having a little fun, even at my expense. As long as nothing gets broken and no one gets hurt, and the project moves forward, a few jokes and tricks are fine with me."

PART FOUR

making history

CHAPTER EIGHTEEN

Nathalie

IT IS SEPTEMBER 7, 2005, NEARLY A YEAR TO THE DAY that I landed in Iquique with Alan Waggoner, Paul Tompkins, Dom Jonak, and Finch, who had picked us up at the airport in his Toyota HiLux and indoctrinated us with the wild and reckless driving habits of rookie roboticists. Those bouncing, harrowing road-warrior-like experiences will forever be lodged in my psyche, especially now at the start of a new school year at Carnegie Mellon, when I catch glimpses of Finch, who is matriculating this month as a Robotics Institute grad student. Finch always waves and grins broadly when I see him on campus, as if he still remembers and is secretly amused by how his "bat out of hell" driving scared us to death.

Since last September, Alan Waggoner has stepped back from the LITA project. Carnegie Mellon will soon announce that Waggoner's MBIC will receive a $13.3 million grant from the National Institutes of Health to establish a research center dedicated to understanding the inner workings of cells and how they can be used in the early detection and treatment of diseases ranging from cancer to cystic fibrosis. Waggoner's

National Technology Center for Networks and Pathways will blend medicine, engineering, and computer science to develop technologies to allow scientists to examine what is going on inside living cells through fluorescent dyes. In addition to Waggoner, Paul Tompkins is absent from LITA. He has received his PhD, and has snagged a job as a researcher and programmer at NASA Ames, joining his roboticist fiancée, Vandi Verma, who has developed Zoë's health monitor.

View of the Compania Minera Punta de Lobos, the site of the biggest open cast mine of common salt in the world. *Courtesy of Alan Waggoner.*

Dom Jonak is still here. He's worked his way up the ladder and is now a key Zoë programmer and caretaker. In 2004, Mike Wagner was Zoë's shadow, the roboticist most in synch with Zoë, but now Jonak has assumed that role. Wagner is still around, but divides his time among Zoë and other projects.

Some important features have been added both to the FI and Zoë over the past year. Zoë now has a plow, an instrument she can lower into the red dust to scrape away layers of soil so that signs of life under the topsoil can be compared with data collected from the surface. And she's also got an automatic spritzer. Heys's design allows the FI to drop down from Zoë's flat fiberglass belly. Then three narrow locustlike arms emerge, from which nozzles are deployed. There's a nozzle for water (and vinegar), a nozzle for the dyes seeking signs of life, and a nozzle for a dye marker spray, so that the ground truth team following Zoë can go to the exact spot where the FI has been testing.

Red Whittaker has not returned to LITA, nor is he expected. Not only has his Red Team rebuilt and significantly improved Sandstorm, but Whittaker himself has also maximized his potential of achieving victory in the upcoming $2 million DARPA return Challenge by giving Sandstorm a partner. H1ghlander is a newer and vastly improved Humvee (actually a Hummer, the commercial offshoot of the military's Humvee) donated to the Red Team after the first DARPA race. Whittaker has not kept his race plan a secret: H1ghlander, newer and faster, equipped with the most advanced technology, will sprint out ahead of the pack on race day, October 15, while Sandstorm, the backup vehicle, will run a more conservative race. They'll be one-two at the finish, for sure. That was the idea. And why not?

After DARPA 2004, sponsors jumped on the Red Team bandwagon and poured in resources, well beyond expectations. On campus, the Red Team got its own building. Red Team paraphernalia was everywhere—jackets, T-shirts, hats—for sale in the CMU bookstore or online at the Red Team Web site. Attending a Red Team event—and there were many of them—was like reliving Texas high school football's *Friday*

Night Lights—cheering fans, reporters and cameras every-where, crews from local and cable channels tailing Whittaker, capturing his every word. The Red Team was bleeding money at the rate of $5,000 a day, an administrator said.

While the Red Team this year has two opportunities to prove itself, for LITA in 2005 there are three OPS scheduled in three different locations. The first OPS is taking place at the original base camp, near the salt mine, Compania Minera Punta de Lobos, where I first saw Zoë in action. Now at Salar Grande, the 2004 start of these OPS, David Wettergreen is feel-ing positive, as if the action this year will finish on a triumphant plateau. The 2004 OPS ended on a much higher note from how it started, and the work his troops have done on Zoë over the previous nine months has helped to straighten out many of the software problems that had plagued them in the 2004 OPS. Dan Villa has graduated and returned to the West Coast.

The PE had been improved; it now works reliably. The executive had been refined so that operator interventions could occur more easily. The navigation system had been sharpened so that it could distinguish between streambeds and more formidable obstacles and made more aggressive so that Zoë would charge forward, climbing hills without hesita-tion. But the most satisfying and exciting aspect is the supple-mentary grant the LITA team has received from NASA for a special and crucial project to be included in the OPS this year. Wettergreen and his colleague Nathalie Cabrol call it "Science on the Fly."

David Wettergreen first met Nathalie Cabrol in 1997 when he was a postdoc research fellow at NASA. Cabrol, whose NASA affiliation is anchored at the SETI (Search for Extra-terrestrial Intelligence) Institute at NASA Ames, was then one of those rare scientists who believed that robots could enhance scientific research in exotic habitats—on Earth and elsewhere in deep space.

Cabrol actually specializes in aqueous environments and arid places where water was once thought to be prevalent.

Nathalie Cabrol on the 2005 NAI/SETI/NASA ARC High Lakes Expedition. *Courtesy of Peter Coppin.*

Water leaves its mark and records its history on rocks. The past presence of water on Mars, which has recently been substantiated, partially by Cabrol, not only provides meaningful implications in the search for extraterrestrial life but also opens up the possibility of human exploration. "We could have water to drink," she smacks her lips, as if this were a rare delicacy.

Slender and chic, Cabrol tends to dress in black and speak in an English laced with awkward phraseology and punctuated by eye-rolling gestures. Cabrol, forty-two, has long been recognized for her expertise, beginning with her special interest in studying ancient high-altitude Martian lakes in an area known as Gusev Crater, which was seen in the very early Viking Mars explorations in the late 1970s. Here is where she believed evidence of water on Mars would be discovered. She wrote her master's thesis on her theories about Gusev in 1985. Not much attention was paid to it in the United States, but Soviet scientists were intrigued. Cabrol, a French woman, became a special consultant to the Soviet Space Agency, then planning its own rover landing on Mars. Gusev was a favored proposed landing site.

"Nobody was working on Gusev at the time, except me. I was invited to come to Moscow [in 1990] to present some of my work. You can't imagine—I was twenty-five, not even a PhD, and here I was in front of the Russian Academy of Science." She received her PhD in 1991 and accepted a postdoctoral fellowship at the Observatory of Paris, Meudon, with a professor who was working in something called "planetology," which, she learned, was a field of study combining astronomy and geology.

Planetology instantly possessed her. She would be, she decided, a planetary geologist. At the time, she was already acquainted with seventy-year-old Edmund Grin, a geologist who explored nearly every mysterious corner and crevice on Earth; he shared her interest in Gusev and supported planetology as her new specialization. Their mutual intellectual pursuits turned into romance and, despite the four-decade age difference, they were married in 1995. Grin would eventually coordinate ground truth for LITA.

Cabrol's one-year fellowship lasted nine years. It was here at the Observatory that she met NASA's Chris McKay, a regular visitor to Meudon and a planetary scientist at NASA Ames. McKay was of the rare breed of scientists who believed that robots could potentially conduct good science in areas inhospitable to humans, such as Mars. At Meudon, McKay looked troubled and distracted. She saw him in the park near the Observatory one day and inquired about his mood. At Ames, he explained, he was trying to design a mission destination for a rover space exploration project, which, at the moment, was stalled because he could not decide on a landing site. Cabrol invited McKay to her lab where a large image of Gusev Crater was lying on her worktable. McKay looked—and he began to smile. Not long after, she and Edmund were invited to NASA and integrated into the SETI Institute.

Subsequently, Cabrol became a leading figure on the science team for Spirit and Opportunity. Through Cabrol's influence, Gusev Crater became a primary target for exploration. Although Cabrol fantasized about personally landing on Mars, she quickly realized that the potential Martian astronauts in the foreseeable future were robots, which is how and why she first became associated with Carnegie Mellon. Robots could discover evidence of the water she suspected existed in Gusev and, at the same time, isolate the rocks here on Earth—the key to the wisdom and secrets of our evolution. "The forces of geology that have shaped the world over the eons talk to us by leaving their memories in what I call the 'book of stone,'" said Cabrol. "Almost all our history is recorded in the rocks. We just have to learn how to decipher them."

Nathalie Cabrol has been involved in robotics since the Nomad project, for which she was lead scientist. In fact, the fossil observed by Nomad in 1997—the same year she met Wettergreen, and the event that paved the way for Spirit and Opportunity, Hyperion and Zoë—was actually discovered by Cabrol.

Finding the fossil in 1997 was significant, but only a very basic beginning—a tantalizing hint of the potential of robots functioning independently in the field and practicing science autonomously. But the problem and the challenge in using a robot as a geologist or biologist, is not, Cabrol realized, what the robot sees as it traverses the terrain, but what it ignores. "The rover only sees through its cameras where we direct it," Cabrol told me. "There is a vast no-man's land in between."

A human scientist walking across the desert or creeping through a jungle, she explained, can take a personal panorama with her eyes every step of the way; she will recognize signs of life, theoretically, or isolate an area that seems ripe for scientific investigation. The rover, however, is handicapped with tunnel vision. It only sees where it is told to look. In 1997, Cabrol, the human scientist, saw the rock with the fossil through Nomad. The rock was retrieved by a human in Chile and then shipped to NASA for tests to confirm Cabrol's suspicions. But robots can and should do more than see for the human. Scientists must find a way to take full advantage of a robot's potential to cover a lot of ground by making it observe the world with a humanlike perspective.

Cabrol always becomes very animated when discussing or explaining the philosophy and objective behind Science-on-the-Fly. She waves her arms and intuits what the rover is supposed to see and think and even say: "We want the rover, as it is traversing across the desert, not to just see, but to be able to understand *what* it sees and then, to say, 'Oh, look!

Here is something that matches my mission objectives. I don't know what it is, I am not at this moment smart enough to know that, but what I am seeing is interesting, so I think I am stopping here and I am calling home to the science team so that they can have a look.' " That is her description of a first step. The robot can't do science, but the robot is aware that it has confronted something that could be valuable to a scientist, so it seeks scientific help.

Then she also envisioned a more advanced scenario. "The smart robot—a robot scientist—might not even call home right away. It would go to the rock that looks interesting, conduct a few tests and get preliminary data, and then call home and say, 'Hey, this is what I found here in the Atacama a few minutes ago. Is this what you are looking for? Do you want me to do more investigating?' " That was a second step in which the robot could actually perform some basic science. But here again, this was a feat that barely scratched the surface of the robot's potential.

"Or even better yet," she explained, her voice becoming higher and her movements more animated. "The robot sees something interesting, so she conducts a preliminary test or two and the results are encouraging. But this time, she does not need to call home. It is too soon in the process to call home. She is more sophisticated. So she conducts more detailed tests, gathers lots more data, and then, finally, she's satisfied that she's done something good, so then and only then does she send it back to home for analysis. That's what a scientist would do walking through the desert, and that's why David Wettergreen and I call it 'Science on the Fly.' "

Cabrol and Wettergreen had wanted to include Science-on-the-Fly in their original LITA proposal, but they feared, as with Whittaker's total autonomy idea, that the project would sound too radical to fund. So they wrote a second proposal for Science-on-the-Fly and submitted it at the same time as the LITA grant. It was not funded. But they revised and resubmitted the proposal the following year, and this time they were successful. (The Science-on-the-Fly funding would eventually extend through 2007 for a total of $800,000.) So in addition to debugging or rewriting Zoë's software programs for the 2005 OPS, the roboticists are unveiling this science autonomy software. If everything goes well on these OPS, Zoë will not only follow directions laid out by the remote science team, Zoë will not only perform autonomous traverses, Zoë will not only deploy her spectrometer and FI, but she will also discover signs of life on her own. She will do autonomous science— Science-on-the-Fly.

CHAPTER NINETEEN

Pirate's Cove

W E ALL GATHER IN THE EVENTSCOPE LAB, WHICH IS located on the second floor of a bland seventies-style building owned by Carnegie Mellon, but detached by a city block from the heart of the campus, and nestled between a Mobil station, a bar and pizza shop, a Chinese restaurant, a Jewish-American Cultural Center, and a Starbucks. EventScope is a virtual-reality software package that helps interface Zoë in the Atacama with the remote science team in Pittsburgh.

The lab, which we are calling the Remote Science Lab, resembles a television studio: The ceilings are painted black, the windows are shrouded, spotlights from track-lighting strips illuminate various work areas, and a glowing, multicolored luminescence flickers across the room from two large projector displays directly in the center of the lab where everyone can see them. The displays indicate, among other things, where the team thinks Zoë initially landed (where roboticists placed Zoë in the field at the start of the OPS) and how far Zoë has traveled this week since they have been working together.

Now, astrophysicist Andy Hock, at the command post, leans forward and raises his fist. He is lean and tall, with hollow cheeks, a jutting unshaven chin, and intense brown eyes shaded with wire-rimmed spectacles. When he announces the start of the 4:00 P.M. meeting by shouting out, "Go science team!" the conversations swirling around him trail off, and the team members quickly return to their laptops and workstations.

The scientists will usually first appear at the lab at 1:00 P.M., settle in at their desks, and gradually begin to examine the data received from the Atacama the previous night. This is usually a quiet, productive period, for there's a wealth of data to review. Even with three hours, they barely skim the surface, eyeballing FI images and panoramas long enough to consider where Zoë might go and what she might do the following day. The data will be studied much more carefully after the OPS and LITA have ended and many papers and journal articles will report the results in staggering detail.

The first science meeting usually begins at 4:00 P.M., as it has just been initiated by Hock. Here the scientists begin to share their observations and insights about the data Zoë sent from the field the day before and start to discuss the science plan for tomorrow. Data from today's activities will start arriving at approximately 8:00 P.M., at which time they will meet again to synthesize it with the 4:00 observations.

At 11:00 P.M., Hock assumes that everyone has examined both data sets and he begins to gently guide the group toward a final plan for Zoë to execute in the Atacama the following morning. They will discuss and debate until midnight. When everything runs smoothly, the science team can begin to shuffle off to their hotel rooms for some badly needed sleep by 1:00 A.M.

But now, after everyone is situated, Hock quickly outlines weather conditions in the Atacama and summarizes Zoë's activities through the first week to today. This precipitates a spontaneous and somewhat erratic discussion that fills the darkened room with a low-level buzz of voices.

For a while, Hock allows the dialogue between the geologists, the biologists, and the roboticists to play out and jell— which is exactly his job as the science team's coordinator—to synthesize information from the many different perspectives represented in the group and to encourage give-and-take so that all points of view are shared. Hock is usually laid-back, as are the members of the science team, but today Hock senses an edge of anxiety pervading the room. His team seems—and he uses a rather unscientific descriptive word to pinpoint their mood—"off."

Although they share common objectives, the scientists perceive Zoë from different orientations. Geologists like James Dohm, from the University of Arizona, are concerned with mapping. They want to know where Zoë is, exactly, and the intimate details of the terrain surrounding her. Biologists, on the other hand, are invigorated by the atmosphere—the weather conditions—surrounding Zoë and, more important, the knowledge yielded by what's above and under the ground beneath her.

The arrangement of the scientists in the room reflects their orientations. Kim Warren-Rhodes, an ecological microbiologist, headquartered at NASA Ames, whose research has focused on hypoliths (algae under [hypo] stone [lith]), and the Carnegie Mellon instrument specialist and biologist who helped build the FI, Shmuel Weinstein, sit side by side beside Hock—directly in front of the displays. On the other side of the room, Dohm, a large man with a booming voice, huddles with his laptop at a table clicking on his terrain maps. Headphones with New Age music isolate him even further.

Jen Piatek of the University of Tennessee, the spectroscopist (she works under Jeff Moersch), also spends most of her day isolated with headphones. It is not surprising, with a row of rivets in her ears and rings dangling from her eyebrows, that she listens to heavy metal—a counterbalance to the orderly, structured science of spectroscopy. In order to measure light and heat, which is essentially what a spectrometer will do, Piatek devotes most of her day to looking "at squiggly lines on paper," the way in which the data are recorded.

The scientists are young. Hock is twenty-eight; Piatek, thirty-two. But the roboticists are younger, serving primarily in a support capacity related to how Zoë is functioning—the meshing of hardware and software. "We are all nerds," says Warren-Rhodes, thirty-seven. "The robotics guys are baby nerds and we are older nerds. But we are all driven by the desire to unravel a complex intellectual puzzle."

In some ways, the roboticists seem more satisfied and fulfilled engaging with the remote science team in the lab than in the field with their peers. Chile has its virtues, especially when you first arrive and are exposed to its stark beauty, but working with the remote science team in Pittsburgh allows Dom Jonak to play volleyball in the evenings, his passion, take a hot shower every night, and partake of the truck food on campus—a respite from trudging along behind the robot and eating its dust.

In addition to serving as consultants to the science team, the roboticists will upload information gathered by Zoë, present it to the scientists, and then at the end of the day send the next plan to the engineering team in the desert. This is the downside of working with the science team, says Jonak. "The scientists are out drinking beer, and we are the last to go home."

Jonak doesn't really know if the scientists are carousing at 1:00 A.M.; he's just joking. But Hock would not be surprised to learn that the remote science team is suffering a collective hangover caused by the results of their work over the past few days. He senses that the scientists are not as happy or as satisfied as they ought to be at this moment, and he tries to help them recognize their accomplishments with a new clarity.

"I know that this is not what we wanted," Hock says. He is referring to the fact that Zoë did not reach Pirate's Cove, as the scientists had yesterday directed her. "And it is frustrating," he continues. "Why do we keep misunderstanding the rover?" he asks, rhetorically, verbalizing what he knows some of his teammates are wondering. This is their second year working with Zoë. Perhaps they feel, Hock theorizes, that they should be in perfect synch with her.

Pirate's Cove is a place they named to give Zoë a specific and verifiable destination and to provide a marker to refer to. But you will not find Pirate's Cove on any map other than the one illuminated on the two large displays in front of the command post directly facing Hock. They called it Pirate's Cove because it seemed foggy and eerie on the DEM—a place, they theorized late one night in the dark lab, where pirates would hang out and ambush unsuspecting ships. The later and the longer the OPS, the more spacey and imaginative the members of the science team become.

But if you look closely enough, and you are in the proper frame of mind after being enclosed in the lab for five days, twelve or fourteen hours nonstop, as are these scientists, you will see that there are areas that actually do resemble pirate places. You will recognize the shape of a three-cornered pirate hat—at a location right where they wanted Zoë to go. There's a silhouette of a pirate with a parrot on his shoulder. They see a raven. They see a scabbard. They even find a place that resembles a pirate's plank, "as in walking a plank," which they name, in honor of the legendary Nobel Prize–winning physicist, the "Max Planck."

"For a couple of days, we were pirate crazy," says Shmuel Weinstein, who went home late one evening with pirates on his mind and returned early the following afternoon with Pirate hats, Pirate jerseys, a Pirate sword—regalia immediately appropriated by Nathalie Cabrol, who prowled the lab, uttering pirate jargon and walking with a charming, exaggerated Pirate swagger. The souvenirs were appropriately baseball-oriented, since Pittsburgh is the authentic Pirate town.

BUT THERE ARE INTERLOPERS in the lab. In fact, there are many interlopers, a few in plain sight and others carefully concealed. The biologists, the spectroscopist, the geologists, the roboticists, even the imaginary pirates, are all being observed, day and night.

Here is Kristen Stubbs, a PhD student at the Robotics Institute. For the past three years, Stubbs has been working on the NASA-funded Anthropocentric Robotics Project, developing a "cognitive model" of how people understand robots and "how a person's mental model of a robot changes over time." Stubbs has observed NASA personnel working on the MER and museum employees interacting with Illah Nourbakhsh's tour-guide robots. Her interest is in exploring human / robot interaction, a field also known as social robotics. She is partnering with Pamela Hinds, an associate professor in management science and engineering at Stanford University, who is in Chile, the interloper on the field team.

NASA's Mars scenario is long and involved and changes periodically with the availability of money and the priorities of the president and Congress. But the long-range goal is based on a close and productive collaboration between humans and robots: Robots will land on Mars or the Moon and work through telepresence or teleoperation with scientists and engineers on Earth. The robots will be paving a way for the arrival of humans, perhaps by constructing shelters or preparing a landing site for a space vehicle. Eventually, humans will arrive to work with the robots, side by side, and finish what the robots have started. From the very beginning, the success of the mission may well hinge on the connection between man and machine—how well they understand each other and how well they work together. So the information collected by Stubbs and Hinds and their conclusions will be crucial to the overall Mars landing plan, once it is put into place.

Kristen Stubbs works very much like an anthropologist in the jungle, conducting ethnographic observations, sketching diagrams of the room and where the participants place themselves. She sits quietly in the lab, writing down conversational exchanges, word for word, and noting conflicts and misunderstandings between the remote science team in Pittsburgh and the field team in Chile, as well as between the scientists and Zoë and the scientists and roboticists at home. In 2004, Stubbs observed the science team for 138 hours. Pamela Hinds logged 241 hours of observation during the OPS in Chile.

In 2004, Stubbs and Hinds discovered many flaws in the attempted collaboration between the roboticists and scientists, beginning with the communication glitches. In a technical report written with David Wettergreen, "Challenges to Grounding in Human-Robot Collaboration: Errors and Miscommunications in Remote Exploration Robotics," Stubbs and Hinds explained how geographical separation limited the context and content of the information being traded from Pittsburgh to Chile, leading to serious misunderstandings and a lack of productivity. Once, for example, the science team received a fluorescent image in which nearly half the field of view was glowing—so strong an indicator of the presence of life that its arrival caused excitement—along with doubts about the accuracy of the data. Had the imager discovered a particularly fruitful specimen or had the camera malfunctioned, they wondered? A day of debating later, the team concluded that the strange glow that had turned them on was actually the glare of sunlight shining underneath the rover.

This lack of contextual information often led to resentment. Because of the importance of measuring for signs of life when moisture and fog were most prevalent, the scientists frequently requested early-morning tests—a request that the field team often resisted. At one point, Stubbs overheard two scientists discussing the lack of early-morning data. One scientist speculated that the field team was lazy; "the rover team didn't want to get up in the morning."

It was true. The field team resisted early-morning OPS, but with reason. The OPS site was a half-hour from base camp, where they were sleeping. Conducting tests at sunrise, say at 6:00 A.M., the field team would have to wake at 4:30 A.M., stumble around in the frigid, moist, sunless desert, pile into their Toyota trucks, and navigate treacherous terrain in the foggy ink of darkness. Arriving at the OPS site, Zoë would have to be powered up, illuminated by the trucks' headlights. All of this without breakfast; the field team could not eat until 9:00 A.M. because of the schedule at the mining camp. Early-morning OPS meant that they missed breakfast entirely.

Context was a small part of a much larger communication problem. The different backgrounds and orientations of scientists and engineers were more relevant. A lack of understanding of each other's priorities—and the ways in which the contrasting groups work—influenced the data being collected and caused annoying frustrations.

In the Atacama in 2004, the scientists repeatedly urged the engineers to install a dew sensor on an instrument that would detect subtle environmental changes and provide important information about the existence of life. The engineers were clueless as to the value of the dew sensor. And besides that, there was another dew sensor on a weather station about a mile away from the base camp. Never mind, however, that the weather station wasn't functioning. And, according to Stubbs and Hinds, "there is no evidence in our data that the rover team told the science team that it was not physically possible for a dew sensor to be mounted on the rover."

Even when the scientists and engineers were in direct communication, they often did not understand each other, even on the most basic levels. Mission "priorities" were completely misinterpreted. Scientists designed each day's plan as a journey and sequence of events, leading to a critical culmination. Highest priority was where they wanted to be at the end of the day. "This confused the engineering team," write Stubbs and Hinds, "because from their perspective, the actions at the end of the plan were least likely to happen" because of the many things that might go wrong in the interim.

You'd think that sooner or later the two teams, composed of incisive, highly educated scientists and engineers, would realize that they were misunderstanding each other and that the misunderstandings were undermining the mission, but through most of the OPS, they remained obliviously unaware. Even the summer workshop at Carnegie Mellon, conducted a few months before OPS, did not significantly enhance the way in which the scientists and engineers understood each other.

"Frequently," began Stubbs, sitting back and laughing at her recollections of the OPS. She is a short, slender academic with rumpled clothes, disheveled blond hair, and a bright electric smile. "The scientists would go off on tangents and fantasize about how much fun it would be if they could be out in the desert on their own with their hammers, smashing rocks and looking under rocks and in between rocks—and digging. It was so hard for them to face the fact that all of this work they love to do, might be done instead by a robot!"

Fantasizing about being in the field rather than cooped up in a dark lab was especially apparent at the start of the LITA project. "Year one, it was driving me crazy not to be able to stick my hand in the sand!" Like many of the scientists, Kim Warren-Rhodes had extensive experience seeking life in the desert, but had never worked with a rover before and was at first unimpressed with the information sent back by the robot. The early data were interesting, she said, but not as good as a scientist on the scene might have been able to gather.

She compares the challenge of finding life on Mars to a "needle in a haystack." Viking in the late 1970s landed on Mars, conducted a few experiments, and found nothing. No signs of life, she says. As she speaks, she is wrapping her hair around her index finger, a persistent nervous habit, and punching the air with her pen, pointing at the enlargement of the DEM hanging on the far wall. "We could have conducted hundreds of Viking experiments on Mars with the same result—nothing." Which does not prove that life does not exist on Mars, only that Viking couldn't find it, perhaps because it remained nearly stationary; a robot or a human would not have been frozen to a single spot. No one expected to find life in the Antarctic either until someone discovered "the last hold-outs, hypoliths."

The challenge is in understanding habitats; what the scientists learn about where life exists in the Atacama will help determine where to look on Mars. Understanding habitats is also the reason for changing science sites; Zoë moved in 2004 from Iquique near the ocean to inland at Angofagasto. This is why robots will be of great value on the Red Planet. "Humans will not easily traverse 200 kilometers away from their home base, while Zoë can cover the distance mandated by the presence of salt, moisture, geologic conditions, or other priorities without endangering human lives—without humans at all!" Zoë was moved by truck in 2004, however, not quite ready to go it on her own.

As a scientist with years of hands-on field experience, Warren-Rhodes has had to make a significant transition in orientation in order to use Zoë effectively. She has given up a measure of control. "I have had to give up my sense of always identifying things visually; I have learned to sit back and let the rover do its work, and then analyze what the rover gives us—not what we get on our own. This has taught me to look at data more carefully and to keep my eyes open for more subtle clues."

Each scientist perceives working with Zoë with a slightly different orientation. "With the rover," says Andy Hock, "you have one uplink and one downlink per day. But in the field, if you lift up three rocks and see what is underneath them, then that is already three uplinks and three downlinks."

Jen Piatek compares the OPS to a site-seeing tour. "We are driving around the Atacama with Zoë, and it is like being on one of those tour buses. You get to see the Eiffel Tower for an hour and you wonder if you have enough time to get to the top and back down before the bus leaves for another destination. Keeping up with the data is like constantly getting on and off the bus. You have ten minutes to look at each piece of data and you need two hours to really give it a good look." This won't be accomplished until after the OPS are over.

As the roboticists bonded with Zoë in the field, the pronouns of reference changed from "it" to "she." Similarly, Kristen Stubbs noted that the scientists gradually began referring to Zoë as "we" in relation to "where we are" in the Atacama and "she" when they found Zoë's actions puzzling, as in, "What is she thinking?" They too were bonding with the machine.

Kim Warren-Rhodes compares being in the field and using a robot to do fieldwork to the differences between riding a bicycle and riding a motorcycle. "On a bike, you can look around and stop easily, while on a motorcycle, you can go farther, see more, and climb hills you might not conquer on a bike. As a team, we have progressed to the motorcycle level, but frankly I am ready to transition to a Lamborghini."

Her generally ambivalent response to the early data in 2003 was due in part to the fact that Alan Waggoner's FI had not been functional. With the FI, the scientists actually have a laboratory on the scene—a vital feature for Mars exploration.

There won't be a lab on Mars, and bringing samples back for testing would lead to serious contamination issues, thus enhancing the FI's value even further.

Stubbs and Hinds identified at least fifty-seven errors of communication in those two weeks in 2004, causing various disconnects and difficulty with data. In only one of five instances were both groups aware that something untoward had occurred, although they rarely knew the origin of error. For most errors of communication, however, only one group—scientists or the field team—was aware that a problem existed. One group, in other words, knew something wasn't right, but was clueless as to what it was, while the other group was clueless that a problem even existed.

For example, the scientists would request a "compass panorama," meaning that the rover should use the SPI to take images in the four cardinal directions. But the roboticists translated "compass" into a 360-degree panorama. The scientists knew that the data they received didn't make sense, but the roboticists also knew that they had followed protocol. It was a mystery.

The field team in 2004 was perplexed and guilty about its inability to get Zoë working and so they tried a little harder—perhaps too hard. If Zoë could not function, they figured, the scientists would not get their data, and without the data, they could not report positive results—a consequence that would endanger future projects and perhaps even the continuation of this one. Thus, the field team did everything humanly possible to make Zoë respond to the scientists' various requests—often leading to situations and actions that were confusing and ridiculous.

If, for example, Zoë would be facing in the wrong direction to carry out an action requested by the scientists, and there was no way she could turn around on her own, "the field team would pick her up and move her into the right direction," says Stubbs. "I don't think this would happen on Mars." Back in the lab examining the data, the scientists could not understand how Zoë had suddenly flipped 180 degrees.

There were times when the scientists acted equally illogically and inconsistently. Trey Smith informed them repeatedly that Zoë's plow was inoperable, but this fact didn't stop the scientists from repeatedly requesting plow maneuvers. Eventually the field team became so annoyed and frustrated with these requests that they responded with a plow maneuver: Stuart Heys dug up the dirt with a putty knife. SPI images were taken, as directed, and the FI was employed. Back at the lab, the scientists were ecstatic.

IF KRISTEN STUBBS is an interloper in the Remote Science Lab, then Geb Thomas, a professor of mechanical and industrial engineering at the University of Iowa, is Big Brother incarnate. Thomas, an engineer with a robotics background, became acquainted with David Wettergreen and Nathalie Cabrol while he too was at NASA Ames on a postdoctorate fellowship. They worked on Nomad together, as well. But as Cabrol and Wettergreen pursued the concept of autonomous science, Thomas had a divergent vision. He believed in their goals, but recognized that neither engineers nor scientists were capitalizing on the opportunity to learn from their rare experiences and maximize the results.

The fancy name for what Thomas does is "Human Factors Specialist." Traditionally, he's the guy that everyone hates. "The field I come from analyzes workers in factories. I am the guy with the clipboard who stands in the background, measuring how long it takes to turn a screw. Then we recommend how to increase factory efficiency. That's my paradigm." Except that now he is working on a NASA-funded project, and he intends to gather enough information to determine how to "get the most and best science from this investment of time and energy."

His work differs from Stubbs's and Hinds's studies because, as ethnographers, they are noninvasive; they are bound to observe and perhaps explain—but not disturb—the system. Thomas, on the other hand, is perfectly willing to be invasive, and he fully intends to improve the ways in which scientists work and maximize the outcome. Thomas questions the traditional ways in which scientific projects like LITA are performed. He points out that the entire LITA project, including Science-on-the-Fly, and even his own research and that of Stubbs's and Hinds's, are, added together, a magnificent demonstration of an idea or a product.

And what do you get in return, he asks, when the demonstration is over? People write papers, which tend to discuss the demonstration and compare it with similar demonstrations. They report results and, with luck, proceed to a new and perhaps related project. Essentially, the scientists assess the science but not the ways in which the science was tested and performed. "The only way to measure the scientists' performance is to count how many papers have been published, and how often the papers are cited in other papers."

Geb Thomas's mission as a human factors specialist is to dig deep into the process and assess the scientific work as it plays out in action. "Human performance is much more complicated to evaluate than machine performance. You know exactly what you expect or hope for from Zoë—but how to understand and improve what humans do? That is what we need to know."

Throughout the OPS in 2003 and 2004, each scientist was "bugged." Every exchange, observation, and interaction was recorded. Now in 2005, the microphones are back, along with seven video cameras implanted in the ceiling to observe the dynamics of the group. And then there's the human factor: Every five minutes, three graduate students working for Thomas, sitting on the periphery of the lab, record interactions—"utterances"—between members of the science team. Thomas is also constantly observing, maintaining an ongoing narrative of the OPS themselves. With all of this data collected, his group of students will devote at least the next six months to making transcripts, counting interactions, and categorizing utterances in order, among other things, to analyze how and why decisions were made and why certain hypotheses were more productive than others.

"I don't think we will win a Nobel Prize with this work," he admits, as he slouches on a chair in the lab, "but I think we will save NASA a lot of money and help scientists maximize their efforts." For example, Thomas and his grad students noticed last year that the scientists frequently, as part of their daily remote plan, requested stereoscopic images from the SPI camera. Panoramas take time and consume many resources, thereby limiting other scientific activities—a key point because, as Thomas and his students also noted, the scientists often bypassed these data once received. Even when they did consult the data from the three-dimensional panoramas, most of their conclusions about the environment in the Atacama came from other sources. So why devote so much time and energy to panoramas? That's the question the scientists will have to answer in the future.

During the 2005 OPS, some members of the team opted to return home and participate in the daily science meetings remotely by telephone. This is inevitable for the future—people will want to be with friends and family—but Thomas points out that no matter how indispensable the scientists seem to be while on the premises, they lose their influence when they leave the cloistered lab space. He recalls, as an example, a recent instance for another project when he recorded a scientist at home making a very important observation during a teleconference. "Lots of folks were sitting there; they were looking at notes, watching the monitors, chatting with one another. No one was paying any attention to what this guy was saying. Two days later, a scientist in the room made the same observation and the entire group was electrified."

Initially a few of the scientists here had reservations about being monitored so carefully, but the transcripts do not name names—people are categorized only by field or profession—and Thomas's students are told to avoid group dynamics when they organize and catalog the utterances and other information. "The last thing I want to do is to produce a psychological profile of the people we are studying," he says.

Inside the lab, it's nothing less than an ethnographic orgy, with the geologists eyeing the biologists and the scientists appraising the engineers and the EventScope staff trying to determine how to improve the system, with Thomas and his students and Kristen Stubbs watching all of them. Not to mention this author and the reporters from throughout the country coming and going for newspapers and magazines.

Stubbs maintains that she benefited from Geb Thomas's presence because his microphones and personnel directed the scientists' attention away from her. "But once Andy snatched my camera and began taking pictures of me so that I could experience what it feels like being watched all of the time. And while he is standing in front of me taking my picture, I am sitting at a desk writing down, 'Andrew is taking pictures of me so that I can understand what it feels like to be constantly observed.'"

THERE'S A FINAL twist to the proceedings: Every Monday through Friday from 3:00 to 5:00 P.M. the entire group—ethnographers, roboticists, biologists, geologists, as well as the EventScope staff and the writers and reporters—is observed, collectively, by high school students in a video conference with the Adler Planetarium and Astronomy Museum in Chicago. Even Thomas is included.

CHAPTER TWENTY

Hardware vs. Software

A FEW DAYS BEFORE THE SCIENCE TEAM ATTEMPTED TO get Zoë to Pirate's Cove, Zoë's power supply suddenly died. The power comes from the connection between Zoë's batteries, which generate energy to her computers, similar to the transformer and power cord you plug into a wall outlet to charge a laptop. The team had a spare power supply, which, when installed, worked only intermittently. Zoë was stopping and starting haphazardly. So Stuart Heys drove the 110 kilometers round-trip to town in the Toyota truck to buy another power supply, which, when installed, also worked only intermittently. Eventually, ingeniously, Heys chopped up an inverter, that device used to provide AC power through the lighter in the Toyotas, and fashioned a makeshift power supply, which worked. So Zoë began to function efficiently—but, alas, once again, only temporarily. Other problems occurred.

Later, when Heys and Mike Wagner returned to the United States, someone inadvertently disconnected the motor for the pan-tilt unit and when he reconnected it incorrectly, it burned up. Heys was in New York, but working on the phone he was able to locate a similar motor in the High Bay and have it shipped to Chile. But no one on the field team at the time knew how to wire the motor into the unit. Perplexed, Heys then ordered a replica of the original unit and sent it down with one of the engineers joining the team. But it too was installed incorrectly. A day later, the PTU replacement followed in the footsteps of its predecessor and burned up. Other mechanical parts went awry, including the plow and the spray devices on the FI. "It was hilarious," Heys said. "One thing after another." But he wasn't laughing. Being the only person capable of dealing with Zoë's mechanics was clearly taking its toll on Heys this year.

I thought back to the way in which the cameras on Zoë were confused in 2004 and how long it had taken Mike Wagner to decipher the puzzle. "Isn't there a blueprint for Zoë?"

"Take Zoë apart," Heys replied, "and me and Jim Teza are the only ones who can put it back together."

For the most part, the differences in orientation between the hardware and software people at the Robotics Institute were subtle, never a major issue, but they were easily recognizable during such stressful periods. Why are code writers allowed to take the rover apart and put it together again, when they don't know what they are doing, Heys asked, rhetorically? "This entire project," he complained, "is based on simulation, and I don't know if that does anybody any good."

This is a viewpoint repeated in private by other engineers—their belief that Zoë and most other robots at Carnegie Mellon exist only as test beds for software. In many respects, their feelings are warranted if we believe that the brain and not the body is the key determinant of autonomous capability and, for lack of better words, "human-ness" or "cognition."

In robotics, no matter how complicated the vehicle or device, the body parts are only as sophisticated as the software—the thinking element—requires or allows. For RoboCup, Manuela Veloso and other researchers gradually eliminated the hardware challenge and went with already manufactured and relatively sophisticated devices like Aibos and Segways. Not that Aibo and Segway robots are ready to go mano-a-mano with a human soccer team (and Aibos will soon be history, anyway), but the mechanics are far enough advanced to accommodate the existing software. An early Asimo, the one I met in 2003 at the American Open, for example, was an exception. Its body was vastly superior—if you are evaluating it on a "humanness" scale—to its cognitive powers. Honda is now attempting to develop software to catch up with the phenomenal hardware test bed it created.

I came to realize the way in which the hardware or the engineering part of robotics was resolved by observing the early planning stages of the Red Team. During his DARPA Challenge money-raising campaign, Red Whittaker enticed some prominent figures from the auto-racing world to join the Red Team, including the famous Indy and NASCAR team owner Chip Ganassi, who attended many Red Team planning meetings.

Ganassi was to take Whittaker out into the desert and teach him the basics of racing, and to contribute a great deal of money to the Red Team effort, but perhaps his most valu-able contribution came during a heated debate over the choice of vehicle for the race. Whittaker and the Red Team were trying to decide whether to design and build a racer from the ground up or purchase and modify an existing vehicle. It was the typical hardware/software, engineer verses AI debate.

Ganassi listened to the debate for some time before standing up and stating the obvious. "Why are you talking about building a new vehicle?" he asked. "There are plenty of vehicles which can do what DARPA is asking you all to do in the desert. We're only talking about going thirty miles per hour. This is not a hardware problem. The issue here is software: The driver and not the vehicle." This was the rationale behind the selection of the 1987 Humvee, which eventually became Sandstorm, and subsequently the addition of H1ghlander. Simultaneous with the LITA OPS, the two Red Team entries competed against one another and twenty-three other robots at the second DARPA Challenge on October 15, 2005.

IN ADDITION TO H1ghlander and Sandstorm, there are big robots like Team TerraMax, a hefty six-wheel-drive Oshkosh truck that is the mainstay for tough transport for the U.S. Marines, and little fellas like Ghostrider, a motorcycle created by graduate student Anthony Levandowski, who quit the University of California at Berkeley two years ago, inspired by the DARPA Challenge, to make a motorcycle that drives itself and to compete in the precedent-setting desert race. There are two sentimental favorites: Team DAD, created by brothers Bruce and Dave Hall, and the Gray Team from New Orleans. They chose to participate just a few weeks after Hurricane Katrina.

It is dawn. The sky in the desert is awakening with a red and blue aura. Nevada is not as dramatic as the Atacama in the early morning, but the event about to be launched makes the morning equally alluring and exhilarating and crackling with suspense. The air vibrates with chill and apprehension, as the robots churn out of the starting gate, one by one, kicking up clouds of the dry dust hanging in the dawn. The robots are launched at five-minute intervals. It takes two hours until they are all on their way onto the course.

The stands are packed with spectators, for that is how to describe the men and few women who made these amazing machines. After staying up day and night for months, investing time, money, and reputations, all they can do is watch and wait, hoping and praying that their dreams and visions of winning the Challenge are fulfilled. I can see Red Whittaker in the stands, peering through binoculars, stretching his neck for a final glimpse of Sandstorm and H1ghlander, surrounded by a slew of his disciples—most of whom are wearing their trademark Red Team baseball caps with matching T-shirts and windbreakers.

In the beginning, H1ghlander, as expected, sets the pace, with Stanley, a modified Volkswagen Toureg, spearheaded by Red Whittaker's former Groundhog colleague Sebastian Thrun, now at Stanford, and Sandstorm a close-behind third. As the race progresses, H1ghlander becomes even more dominant—churning up the terrain and pulling away from Stanley and Sandstorm, as if they are in another league or another race.

This is the Whittaker plan—H1ghlander is to be the rabbit, looking to blow out its competitors, with Sandstorm as a safety valve just in case something untoward and unexpected happens. DARPA is monitoring the race with helicopters, which send back video of the action for the spectators to wit-

ness on monitors and cheer inside a large headquarters tent—
or to moan in frustration and despair. There are also DARPA

H1ghlander wearing a medal at the second DARPA Grand Challenge, 2005.
Courtesy of Evan Tahler, Red Team, Carnegie Mellon University.

chase vehicles tagging along behind each robot, ready to pull
the plug if any get out of control. Team DAD is one of the
first contenders to falter when its laser suddenly flies off its
roof. The Team DAD chase car monitors quickly terminate it.

With nearly one-fourth of the race behind it and with
many of the major course hurdles (gates, ravines, etc.) neatly
navigated, H1ghlander is gaining power and speed. At around
mile 35, it literary rockets down a hill, clearly a confident pace-
setter, accelerating even faster in the straightaway. Then it
begins climbing the next hill when, suddenly, inexplicably,
something happens, something mysterious and illogical. It is

as if H1ghlander has run out of fuel or popped out of gear, or lost its nerve or will. The robot is acting like it is hypnotized, in a trance. H1ghlander rolls to a stop right in the middle of the roadway, and then, slowly and sickeningly, slides backward down the hill it had just almost climbed.

Sandstorm with flag at the second DARPA Grand Challenge, October 8, 2005.
Courtesy of Evan Tahler, Red Team, Carnegie Mellon University.

It is a pitiful sight, and it seems to take forever for H1gh-lander to awaken from its trance. But eventually H1ghlander, now dust-soaked and clearly wounded, gradually recovers from its dreamlike state and begins moving forward again, slowly. But it is not and would never be the robot it was at the starting line. Perhaps half of its horsepower and all of its inner spirit have now dissipated.

Courageously, H1ghlander claws its way back up the hill, wandering drunkenly on and off the road, fighting to the top. Somehow, its cutting-edge gimbal got turned around, and it is now hanging lopsided, like a hat on the roof of the Hummer. H1ghlander recovers some energy on the following level part of the course, but now, gradually, Stanley recognizes its edge and moves forward. Sandstorm is also closing the gap. H1ghlander actually manages to hold on to the lead until the 100+ mile marker, and then Stanley passes it.

Back in the tents at the TV monitors, the Red Team is collectively speechless. Now they can only hope for a miraculous H1ghlander reawakening—or a misstep by Stanley. At the announcement that Stanley passed H1ghlander, Sebastian Thrun stands up and begins to dance. After driving six hours and fifty-three minutes, Stanley crosses the finish line in the lead position, averaging 19 mph. Thrun is still dancing.

Sandstorm, the reliable pioneer of the first DARPA race, follows across the finish line seven minutes later. H1ghlander staggers to the finish line twenty-one minutes after Stanley, followed closely by the Gray Team from New Orleans. Team TerraMax, the gigantic Marine Corps warrior truck, finishes fifth. All of the other robots die on the way during the 132-mile obstacle course.

Interviewed soon thereafter on PBS, a deflated Whittaker, his Red Team cap slightly askew, seems stunned. "The engine punked on us," he says, quietly. "The only thing that was short was not having a top end speed on the gas pedal. Except for that, we are in like a charm."

Months later, members of the Red Team still were unable to come up with an answer to why H1ghlander tanked. But Carnegie Mellon PhD Michael Montemerlo, who went with Thrun to Stanford to work on Stanley, put the race and H1ghlander's problems in perspective. He said that Stanford's and Carnegie Mellon's software was different to a certain extent, but certainly comparable.

"Winning and losing this race had little to do with software; it was a mechanical problem. A hardware problem. And it was bad luck for Carnegie Mellon. Robotics, you know, is all so terribly unpredictable."

THE LITA FOLKS, however, were not so fortunate as to have a backup Zoë in the High Bay or in the desert as the Red Team had with Sandstorm. Stuart Heys was the next best thing, however, although by the time the 2005 OPS were half over, Heys's nerves were shot—and the worse was yet to come. Zoë was shipped by tractor-trailer truck from the salt mine base camp near Iquique to an inland site near Angofagasto, where the next base camp would be located. Heys secured Zoë as best he could.

The truck took off toward Angofagasto, as did Heys in one of the Toyotas. Mike Wagner and Dave Pane were with him. Hours later, they rendezvoused with the truck at the new site. I have talked with Heys, Wagner, and Pane, all of whom were on the scene the moment the rear doors of the truck swung open and they could see what had happened to Zoë. "Zoë was right in the back, so as soon as we opened the door we saw it immediately," Heys says. Zoë's rear axle was in two pieces. It was irreparable—"a heap of mangled metal."

They waited a while—nearly half a day—before they contacted Wettergreen by satellite phone. They wanted to have a plan before they broke the news to their boss. Fortunately, Dr. Guillermo Chong Diaz of the Universidad Catolica del Norte is a partner. Throughout the LITA project, any time help or support was required in Chile, Guillermo Diaz, a renowned Chilean geologist, responded. Dr. Diaz has been a key colleague ever since 1997 when Nathalie Cabrol spotted the rock with the fossil in the Atacama through Nomad's cameras. Guillermo Diaz was the scientist who rushed out to the desert to retrieve the fossilized algae. Diaz is also an influential geologic consultant for mining companies, which is how Wettergreen came to arrange food service at Compania Minera Punta.

With Guillermo Diaz's support, the Universidad Catolica del Norte responded with quick and gracious hospitality. They provided tools and a comfortable garage area for Heys and the other members of the field team to work. Fortunately, they soon discovered that only the axle was ruined; somehow the computers and Zoë's other vital innards had mostly weathered the bumps and survived. They would learn later that the FI had been damaged, and even though it was operable, it never again functioned with the clarity and efficiency it had displayed prior to the accident. But, at this point, what to do about the broken axle was the major concern.

There simply wasn't time to make a new one—a six-week project at best. Heys knew this because, over the winter, he had fabricated new axles for Zoë, a thought that made him realize that the original axles were still in the High Bay. The old axle would have to be adapted to the other new parts he had fabricated—but he could do it. He called ahead to get the work started, and then flew to Pittsburgh to finish the job himself. In the end, Zoë could continue. It would never be the same robot again, but the mission, at least, could be completed. The OPS suffered only a relatively minor delay.

CHAPTER TWENTY-ONE

In the Field

On the first attempt to get to pirate's cove, the scientists gave Zoë too many jobs to do, Andy Hock points out as the science team meeting continues. With two previous years of OPS behind them, plus nearly a week of this final year, the scientists had expected to be in synch with Zoë, but their appetite and enthusiasm for data far exceed Zoë's capabilities, Hock says, especially considering the difficult conditions in the desert this year.

It was very cold in 2005—one of the chilliest winters the Atacama had experienced in many decades. The wind was sometimes so vigorous that the OPS were delayed or postponed for hours—and the roboticists had to prioritize Zoë's tasks for the scientists, while carefully safeguarding Zoë and themselves.

But Zoë proved to be a persistent and durable robot. Every day Zoë worked its way across the desert, its electric motors whining relentlessly, its knobby tire wheels tattooing the terrain with lines of sculpted imprints—despite the ongoing power supply problem. The robot wranglers like

Chris Williams and Stuart Heys tailing after Zoë now had to hustle to keep up with it, since the speed had increased significantly. And in 2005, they were giving Zoë more space than ever before, not as much as Red Whittaker might have liked, but a kilometer distance as Wettergreen had "hoped."

"Poor Zoë," Kim Warren-Rhodes said one day. "She is working very hard and stressing out."

"But she is sending back some incredible images," says Shmuel Weinstein. The panoramas they are stitching together are clearer and more extensive than ever before, and the images recorded by the FI after the dyes are sprayed light up their computer displays with vivid colors, demonstrating the existence of many different life-forms. But the scientists want more, and they think they need more. They know that the more abundant the science data, especially of this high quality, the more they can help Wettergreen and Cabrol make their case to extend the funding for LITA for another year—or three years.

Wettergreen and Cabrol have recently announced that the last official LITA event for the scientists will take place the first week in January 2006 at the actual scenes of the crime—Zoë's landing sites in the Atacama, beginning with the Salar Grande, near the salt mine, where Zoë is at this exact moment. Visiting the Atacama will be a reward to the scientists, a way to confirm all of their scientific speculation from 2004 and 2005, to see how accurate their forecasts have been. This is a magnificent gift from Cabrol and Wettergreen, they acknowledge. Throughout the entire LITA campaign they have been talking incessantly about actually getting into the field and digging into the dirt, lifting and fingering the rocks they have heretofore only been able to see through a camera lens. Now their wishes will be fulfilled. Soon they will have their chance. They are ecstatic.

THE SECOND TIME the science team tries to get Zoë to Pirate's Cove, they don't do much better than on the first; Zoë's long day of traverses and data gathering leaves it far short of the destination. They are disappointed, again, but this time Andy Hock recognizes an aspect of the previous two days that the scientists are overlooking. By not reaching Pirate's Cove, he tells them that day at the 4:00 P.M. science meeting, they had achieved something significantly more important.

To this point in the OPS, the scientists have been wholly consumed by the task of pinpointing Zoë's location, sending it on long traverses and ordering up full FI sequences. They hadn't allotted time for the autonomous science software to come into play. But for their second attempt to get to Pirate's Cove yesterday, they budgeted time for Zoë to look around on her own. And she did.

Remember that Zoë can't take in the sights like a human being; rather, the new autonomous software allows her to do transects, meaning that she scours the terrain in thirty-meter segments, zipping along in vertical and horizontal lines. All the while she studies the density of the terrain, periodically pausing and moving back and forth almost imperceptibly as she contemplates the color and the shape of the rocks. She reminds me of an animal sniffing for scent. When the vibes seem right—there's something there!—she lowers the FI from her belly and directs the dye spritzer to test for chlorophyll.

There's a slight, hissing whoosh as the nozzle shoots out the dye. The presence of chlorophyll—bacteria are present—is confirmed, so Zoë moves to the next level of science.

Now Zoë signals for an entire FI sequence. The nozzles, which have been retracted, are lowered down the belly of the beast one more time. In twenty minutes, after much clicking, flashing, and spritzing, the results are in. Later in the day, the data will be sent back to the science team for a thorough examination. This is very significant, as Andy Hock has been hinting to his troops that afternoon. But they have become so obsessed over their quest to reach their final destination that they are oblivious to this event that occurred along the way.

"In the long run it doesn't matter that she didn't make it all the way to Pirate's Cove. What happened was better," says Hock. "The frustration that we feel should be more than offset by the fact that Zoë came back with biology results—a totally unexpected development."

While Hock is speaking, Nathalie Cabrol moves to the center of the room, in front of Hock. "Unexpected," says Cabrol, interrupting, "but also what we wanted."

She pauses to survey the room, nodding and shaking her head back and forth, as if confused. "I am surprised," says Cabrol to the scientists in the lab, "because I was figuring to hear more excitement from you all." She pauses again, as the scientists stare at her with curiosity. They still don't seem to realize what has just occurred.

"Forget Pirate's Cove," Cabrol continues. "Realize the significance of last night." She smiles and lowers her voice an octave for emphasis, so everyone will listen more carefully. "This is the first time for 'Science on the Fly'—a rover finding life on its own," she says, finally.

Cabrol pauses to allow her observation to sink in. "You made history," she says. "Zoë made history. No scientist directed it or planned it. Zoë saw something interesting. She stopped. She analyzed its signal and then initiated an entire set of experiments, on her own. She found signs of life, all by herself, without human intervention. She's the first robot scientist—the first robot to do autonomous science."

CHAPTER TWENTY-TWO

The Barest Beginning

Sitting in the remote science lab that day, few of the scientists and roboticists feel the triumph and elation Cabrol is attempting to communicate. They are pleased that Zoë actually performed a scientific experiment, autonomously, on their watch, and also relieved that their inability to get Zoë to Pirate's Cove—what they perceived as failure—has been minimized. Although using the science autonomy software was nearly an afterthought when they sent the science plan to the field the previous night, they can now acknowledge that something better than anything that might have occurred at Pirate's Cove has taken place instead.

But in retrospect, this is the fifth day of "sol" (a Martian day, which is forty minutes longer than an Earth day) of the first of three OPS, and they expect that this is only the beginning of a series of technological and scientific achievements— the culmination of three years of hard work and planning and much glorious dreaming. They don't know at the moment what is going to happen to Zoë next, that Zoë, that magnificent but fragile mechanical marvel, will begin to slowly disintegrate as the many kilometers of wear-and-tear traverses in the Atacama, and the prior testing on Flagstaff Hill and in the High Bay, combined with the sometimes slipshod treatment it received, begin to take their toll.

At the moment they are expecting much more out of these final field operations—a buzz, an ovation, a singular moment of indescribable elation—than they will get in the end, at least as long as they remain in this dark, dreary lab, isolated from fresh air and sunlight and attempting to intellectually connect with a mass of plastic and metal, nearly halfway across the world.

As it turns out, Zoë, limping and struggling, will perform autonomous science at least one more time before the end of the third OPS. Zoë will also traverse, autonomously, long distances—75 traverses of a single command, some as long as 5 kilometers—and discover interesting vegetation. Signs of life will explode in bright and vivid colors on their computer displays as Zoë sends back more and more images—signs of life captured by the FI. And later, the scientists will feel even closer to Zoë because of a program Dom Jonak has made available allowing them to see—literally—what Zoë can see as she traverses the desert.

This was almost, although not quite, as good as being there with Zoë, which was their dream, as they repeatedly stated and fantasized in the lab over the course of three years of trying to do science through a robot: To get to the field, finally, to experience the intense sun and the wind-whipped desert air as Zoë and her field team did; to pick up the rocks that Zoë captured in her panoramas and feel them in their hands, crack them open with their hammers, and to find out firsthand how close they had come to understanding the desert through a robot's perception.

Geb Thomas, the persistent, inquisitive human factors researcher is ready when those dreams are fulfilled and the scientists' boots finally touch the ground in the Atacama for the first time in early January 2006. Thomas, as it turns out, is not in any way prepared for what he is about to discover.

Thomas is a bearded researcher with a gentle voice, an easy laugh, and a friendly and ingratiating manner. To prepare for this final segment of his study, Thomas and an associate had gone to the Atacama in advance of the science team, visiting all of the sites Zoë traversed, which is exactly where the scientists would go in January in order to see what they had imagined and deduced from the Atacama snapshots sent back by Zoë. He then composed a series of true-and-false questions related to each site.

"I had been working under the presumption that once we got the scientists to a particular place in the Atacama, and they looked around at the same spot they had viewed through Zoë, that they would all agree on what they saw." Thomas laughed at the memory and the surprise of the experience. "But that wasn't always the case." Even for the easy series of true-and-false questions posed to the scientists by Thomas and his associate, there were many differences. This was especially surprising because Thomas had lifted many of the phrases and descriptions he used for the questions verbatim from the scientists' own words and observations, written or quoted in the science summaries recorded during the OPS in the lab.

For example, based on observations through Zoë's panoramas, the scientists frequently described what they perceived as many white rocks scattered throughout the desert. The specific phrase they used, says Thomas, was "a lot, a lot of white rocks." But when they got to the site they had described and

THE BAREST BEGINNING · 273

looked around and then were asked the true-or-false question concerning "a lot of white rocks," some of the scientists said "yes" and others said "no" because, Thomas laughed again, "they couldn't agree on the definition of 'a lot.'"

"So even though they had, while in the lab, described many of these areas in the Atacama as having 'a lot of white rocks,' once they were in the desert and on the scene, they debated the meaning of 'a lot,' but not the fact that there were white rocks. Did I get that right?" I asked.

"Yes—and no!" Thomas replied. "Yes, they couldn't agree on the meaning of 'a lot.' But they also couldn't agree over what 'white' was."

Now Thomas was laughing even harder. "Here's what really blew me away," he said. "We came to a 'Salar' [pronounced sol-are]. This is a salt bed where all the water had long ago drained down to the lowest spot and then evaporated, leaving salt and pillars of salt. The pillars were twenty to forty centimeters high and looked very funky. They shot up out of the ground. They were very solid; you couldn't drive a car over them. Dust from the volcanic hills surrounding the Salar had blown over time in this direction and covered up some of the salt on the ground and the pillars. So there were vast expanses of white salt with red dust coatings over it."

Judging by what they saw in the panoramas and fortified by spectrometer readings, the scientists had described the place as volcanic in the lab, although they weren't positive. They needed to see it firsthand to be sure. "So when we get there and we walk into it, they first can't agree if it is volcanic or a Salar." Is it volcanic because there is volcanic dust on the salt, or is it a Salar because there's salt under the volcanic dust? Are the rocks white because they are salt—or red because the white rock is covered in red dust? "The questions and the debate went on and on," said Thomas. Then, to make matters even more complicated, "they couldn't agree on what counts as a rock and what counts as sediment." In other words, was the salt actually rock—or salt? Was salt . . . rock?

The disagreements weren't only between the biologists and geologists, Thomas said, although, as expected, it often did split between specialties, but the geologists frequently also disagreed with one another. "One of the geologists kept insisting, 'I need to take samples and bring them back to the lab, to really know what I am talking about,'" which, said Thomas, indicates at least to this particular geologist here that truth lives in the lab, even when you are standing there and looking at things, firsthand—which is exactly what they had been wanting to do from the start.

By bringing the scientists to the Atacama, the LITA team was able to confirm the fact that most of the observations made by the science team in the lab were accurate—on the money. At the same time, this trip to the Atacama has led Geb Thomas to rethink the ways in which he evaluates such missions and the problems and challenges he must think about confronting in the future.

"Up until now, I felt that the way to measure error in a robot mission was the difference between what you saw with the robot from the control room compared to what you saw when you were standing there where the robot had been working. That was my fundamental logic, but now I have my doubts about what my fundamental logic should be." These findings have led to an ongoing and increasingly complicated discussion with his students and colleagues over the meaning of "truth" and "accuracy." He feels that his measurement guidelines remain valid most of the time, but no longer is he confident that they are valid all the time. And he is concerned about the future direction of his research, specifically, and science autonomy generally. "How are we ever going to get the scientists to tell the engineers what they need, so the engineers can build the right system, if the scientists can't agree on such surprisingly basic levels? The situation," said Geb Thomas, "is more than a little fuzzy."

"Fuzzy" may be a good word to describe the results of the entire LITA project, the other projects and ideas portrayed in this book, as well as the field of robotics generally. Something special, something important and significant seems to be happening in the robotics world and at Carnegie Mellon University's Robotics Institute, but exactly *what* that is and how long or how bumpy the developmental road might be until that "what" is clarified is an up-in-the-air question. Clearly, the potential of robots has been anchored more in reality over the past three years than at any time over the preceding fifty years, especially early on when some people feared that robots might take over the world. That nightmare might happen in the far-off future—robots, potentially, replacing humans as the integral force of the world—but at least for now and the foreseeable future, most people understand how far away a world dominated by robots is from today.

Even the robots that are here today are far away—decades, at least—from becoming a contributing element in our society. Take Groundhog, for example. Its navigation and mine-mapping software have been perfected since it lost its way in the Mathies Mine, and in subsequent tests at Mathies, Groundhog found its way in and out of both portals, repeatedly, without a glitch. Groundhog may well be retired fairly soon, according to the young roboticist who helped build and perfect it, Aaron Morris, to be replaced by a sleeker, smaller, faster subterranean rover called Cave Crawler.

But since the Quecreek mine disaster that precipitated Red Whittaker's campaign to create a mine-mapping robot, many miners have died in accidents in West Virginia and elsewhere around the world, most significantly at the Sago Mine, where twelve miners perished. And in no case have robots for mapping or rescuing been deployed. One could make a case that mine owners have not geared up for the robotic world by investing time and money. And that is true, in a way. But the bigger and more realistic truth is that despite Groundhog's advances and the upcoming Cave Crawler, technology has not progressed far enough. No robot on Earth today could have saved the men who died at Sago. The evolution of technology and further significant funding by government agencies like DARPA will be necessary before such a rescue machine can become reality.

But now that the DARPA Challenge is over, Aaron Morris reports that even though the subterranean robotics group at Carnegie Mellon is smaller than many of the other project groups, some of the graduate and PhD students attached to the Red Team have started to return to or inquire about mine-mapping opportunities. "They are looking for a future direction," Morris says. "They are saying, 'So what are we going to do with our lives now?' "

Of all of the young roboticists I have been able to work with and observe at Carnegie Mellon, Aaron Morris seems to be the most enlightened and motivated by the impact of the institution, perhaps because of where he grew up. "Being here with all of the PhD faculty and staff, working on their level, side by side, flying to conferences and giving speeches, has been an empowering experience. And learning about robotics from such a diverse group has been awesome. In the same room, you can have an artist, an engineer, an economist, and an accountant, and we all work together and contribute to one another's projects! What could be more desirable?"

Morris has also become a skillful salsa dancer, so good that he travels around the community to give salsa demonstrations. "West Virginia is very isolated. You are in the mountains with families who have not been anywhere else. Here I have done things that I never imagined I would ever do. In West Virginia, if I told people that I would some day work with robots and do salsa dancing, my relatives, neighbors, even my father, would have looked at me like I was out of my mind."

Although many people may have believed that Red Whittaker was out of this mind for imagining he could raise millions and create a robot that could navigate on its own across the desert, he proved them dead wrong—sort of. The fact that five of the twenty-five robots in the DARPA Challenge actually finished the race is a great achievement. Whittaker's robots finishing behind Stanford's Stanley only barely tarnishes his triumph. But the impact of the Challenge—what was accomplished and what is going to happen next—is also kind of fuzzy.

Remember that when the Challenge was announced in 2003, it was supposed to be a 30-mile-per-hour dash across the desert from Los Angeles to Las Vegas, a total of 250 miles. The achievement—132 miles, averaging 19 miles per hour, although significant, is, as David Wettergreen might put it, a definite "fall-back position." But still, the technology developed in order to make the dash will eventually be used in a myriad of applications for the military, and by automobile companies interested in enhancing the safety of their products. And Whittaker, who has turned entrepreneurship ("technological swashbuckling") into a high art, is today deeply involved in a start-up enterprise located on the site of an abandoned steel mill in Pittsburgh, a research and developmental think tank he is calling Robot City. There he will undoubtedly retool Sandstorm and H1ghlander for the next Challenge, recently announced by the Pentagon. The new competition will be carried out in a mock urban area. Robots will be required to obey traffic laws while merging into traffic, as well as negotiating traffic circles, busy intersections, and obstacles. The event is scheduled for November 3, 2007.

But this is the rhythm and the formula for progress in robotics: Dream an impossible dream, reach as far as you can toward achieving that dream, and then, with the realization that the dream cannot come true, at least at this moment in time, readjust your goals accordingly and make your compromises count. Such is the way in which Manuela Veloso positioned her RoboCup scheme.

Veloso and her Sony associate Hiroke Kitano realized that a team of robots defeating the World Cup champions in soccer would not soon occur, if ever. But it was a dramatic notion that could and eventually did capture the imagination of her colleagues and the general public and energize what was then a sluggish robotics movement. Without Veloso and Kitano's vision, the DARPA Challenge would probably not have happened. Few other elements fuel progress more than competition among a group of highly charged, egocentric young men (mostly). It is interesting that two of the three great visionaries of the robotics world captured in this book, however, are women: Manuela Veloso and Nathalie Cabrol.

Zoë's history-making Science-on-the-Fly achievement, as significant as it was to Cabrol, was in some ways a fallback accomplishment to David Wettergreen. "There's quite a bit more to science autonomy than just following up on an image of a rock and doing a series of FI sequences. My dream for science autonomy is that robots will actually be like graduate student scientists. Robots won't need to go back and report to humans and wait for directions. They will think about what they see and generate a hypothesis based upon their own research and analysis. That's my vision."

The scope and breadth of vision pervades the robotics world at Carnegie Mellon. Heroes in science and technology are those men and women who embrace hard work and manifest the ability to see with clarity into the distant future, to believe in themselves and in their dreams and, most important, to never give them up.

On the day Zoë made history, I asked Nathalie Cabrol if at that moment she was feeling a sense of pride for all she had accomplished.

"It is not a matter of being proud; it is a matter of having a vision, sticking to it, and, in the process, making it work. I don't have time to look in the mirror and say to myself, 'You did it! You did it!' because this is only the beginning. I intend to do it again and again, and each time, make it a little better and even more successful than the time before."

"You are a relentless person," I said.

"This is good science," she told me. "Science is one obstacle after another, one wall after another, walls some people thought would stand forever that are getting pushed. Those walls will soon collapse; all the walls will go down, in time. What we have done today is significant, but in the big picture, this is only the barest beginning."

acknowledgments

THANKS TO ALL OF THE CHARACTERS WHO APPEAR in this book—human and robot—for your cooperation, patience, and hospitality. I appreciate being allowed into your exciting and impressive world.

Thanks also to Peter Coppin, Angela Wagner, Dimitrios Apostolopoulos, Omead Amidi, Eben Meyers, Sanjiv Singh, Christopher Atkeson, Sarah Keisler, Al Rizzi, Greg Armstrong, Takeo Kanade, Geoff Gordon, Nancy Pollard, Ralph Hollis, James Morris, Howie Choset, Devin Balkcom, Rosemary Emery, Todd Simonds, Chuck Whittaker, Kathy Whittaker, Deborah Tobin, Michele Gittleman, Eric Close, Alonzo Kelly, George Stetten, Ruth Gaus, Steve Stancliff, Brenda Copeland, David Sobel, Rich (Ringo) Noel, Ron Dahl, Byron Spice, and Bruce Steele.

A special thanks to Chuck Thorpe and Anne Watzman, who made my ongoing access into the Robotics Institute possible.

And last but not least, thanks to Patricia Park for reading early drafts and her encouraging observations; to my editor, Amy Cherry, for her flexibility and her astute critiques; and to my agent, Andrew Blauner, for his steadfast and faithful support.

Portions of this book were published in *Salon* ("Bend It like Robo-Beckham," June 11, 2003) and *Pittsburgh Magazine* ("Genius at Work," March 2004, pp. 48–51).

notes

p. 7   "to enable future space missions": ASTEP Web site: http://ranier.hq.nasa.gov/astep/astep.html.

p. 11   "There was a time": Bruce Steele, "The Red Zone," *Pittsburgh Magazine*, October 1989, p. 36.

p. 14   "Roboburgh": *The Wall Street Journal*, Tuesday, November 23, 1999, p. 19.

p. 53   "It's the goal": Author conversation with graduate student Steve Stancliff. July 3, 2001.

p. 62   "A master-slave relationship": http://vesuvius.jsc.nasa.gov/er_er/html/robonaut/telepresence.html.

p. 66   a report: Report by the National Council for Research on Women (2002), www.ncrw.org.

p.104   article about Wallace: Clive Thompson, "Approximating Life," *New York Times Magazine*, July 7, 2002, p. 34.

p. 117   Forbes.com survey: www.forbes.com/2001/05/10/singles.html.

p. 124   article about Whittaker: "The Red Zone," *Pittsburgh Magazine*, October 1989, p. 36.

p. 140   "shakedown artists": Joe Hooper, "Special Section, DARPA Grand Challenge," www.popsci.com, June 2004, p. 1.

p. 150   A National Council on Aging Study: *Science Magazine*,
         October 15, 2004, pp. 503–7.

p. 244   technical report: Kristen Stubbs, Pamela Hinds, and David
         Wettergreen, "Challenges to Grounding in Human-Robot
         Collaboration: Errors and Miscommunications in Remote
         Exploration Robotics," Tech. report number CMU-RI-TR-
         06-31, Robotics Institute, Carnegie Mellon University, July
         2006.